A Taste for Change

THE ECOLOGICAL TRANSITION AS A WAY TO HAPPINESS

Gaël Giraud, SJ, and Carlo Petrini
in dialogue with Stefano Arduini

Foreword by Pope Francis

Translated by Carla Ranicki

I0869434

Georgetown University Press / Washington, DC

© 2025 Georgetown University Press. All rights reserved. No part of this book may be reproduced or utilized in any form or by any means, electronic or mechanical, including photocopying and recording, or by any information storage and retrieval system, without permission in writing from the publisher.

The publisher is not responsible for third-party websites or their content. URL links were active at time of publication.

Original title:
Il gusto di cambiare. La transizione ecologica come via per la felicità

© 2023 SLOW FOOD EDITORE S.R.L.
www.slowfoodeditore.it

© 2023 – Dicastero per la Comunicazione – Libreria Editrice Vaticana

Cataloging-in-Publication data is on file with the Library of Congress.

978-1-64712-595-0 (hardcover)
978-1-64712-596-7 (paperback)
978-1-64712-597-4 (ebook)

♾ This paper meets the requirements of ANSI/NISO Z39.48-1992 (Permanence of Paper).

26 25 9 8 7 6 5 4 3 2 First printing

Printed in the United States of America

Cover design by Brad Norr

A Taste for Change

CONTENTS

FOREWORD

The good that appears as beautiful carries with it the reason why it must be done. This is the first thought that arose for me after reading this beautiful dialogue between Carlo Petrini, whom I have known and esteemed for years, a gastronome and activist known all over the world, and Gaël Giraud, a Jesuit economist whose contributions I have recently appreciated in *La Civiltà Cattolica*, which has published his articles on economics, finance and climate change.

Why this connection? Because reading this text generated in me a real "taste" of the beautiful and the good, that is, a taste of hope, of authenticity, of the future. What the two authors bring forward in this exchange is a sort of "critical narration" with respect to the global situation: On the one hand, they elaborate a reasoned and compelling analysis of the economic-food model in which we are immersed, which, to borrow a writer's famous definition, "knows the price of everything and the value of nothing"; on the other, they propose several constructive examples, established experiences, singular stories of care for the common good and the commons that open the reader's mind to a view of goodness and trust in our time. Criticism of what is wrong, stories of positive situations: one with the other, not one without the other.

I would like to highlight a significant fact: In these pages, Petrini and Giraud, one an activist in his 70s, the other an economics professor in his 50s, that is, two adults, find in the new generations established reasons for trust and hope. Usually we adults complain about young people, indeed we repeat that the "past" times were certainly better than this troubled present, and that those who come after us are squandering our achievements. Instead, we must admit with sincerity that it is the young people who embody the change we all objectively need. It is they who are asking us, in various parts of the world, to change. Change our lifestyle, so predatory towards the environment. Change our relationship with the Earth's resources, which are not infinite. Change our attitude towards them, the new generations, from whom we are stealing the future. And they are not only asking us, they are doing it: taking to the streets, demonstrating their dissent from an economic system that is unfair to the poor and an enemy of the environment and seeking new ways forward. And they are doing it starting from the everyday: making responsible choices about food, transport, consumption.

Young people are educating us on this! They are choosing to consume less and experience interpersonal relationships more; they are careful to buy objects produced following strict rules of environmental and social respect; they are imaginative in using collective or less-polluting means of transport. For me, seeing these behaviors spreading to become common practice is cause for consolation and confidence. Petrini and Giraud often refer to youth movements that, in different parts of the world, advance the demands of climate justice and social justice: The two aspects must be kept together, always.

The two authors point to operational paths for sustainable economic development and criticize the concept of prosperity that is in vogue today. The one according to which GDP is an idol to which every aspect of living together is sacrificed: respect for the environment, respect for rights, respect for human dignity. I was very impressed that Gaël Giraud reconstructed the way in which historically GDP has established itself as the only parameter to judge the health of a nation's economy. He states that this happened during the Nazi era and that the reference point was the arms industry: GDP has a "war" origin, we could say. So much so that this is why the work of housewives has never been counted: because their efforts do not serve the war. More proof of how urgent it is to get rid of this economistic perspective, which seems to despise the human side of the economy, sacrificing it on the altar of profit as the absolute yardstick.

The nature of this book is also doubly interesting. First, because it takes place in the form of a dialogue. This is something I think it is important to emphasize. It is confrontation that enriches us, not standing firm on our positions. It is conversation that becomes an opportunity for growth, not fundamentalism, which bars the way to novelty. It is debate that matures us, not the hermetic certainty that we are always "in the right." Even and especially when we talk about the search for truth. Blessed Pierre Claverie, Bishop of Oran and martyr, said: "You do not possess the truth, and I need the truth of others." Let me add: The Christian knows that he does not conquer the truth, but rather it is he who is "conquered" by the Truth, which is Christ himself. This is why I strongly believe that the practice of dialogue, confrontation and encounter is today what most urgently needs to be taught to the new generations, starting from childhood, so as not to foster the construction of personalities locked in the narrowness of their own convictions.

Secondly, the two interlocutors—wisely stimulated by the editor—represent different points of view and cultural origins: Carlo Petrini, who defines himself as an agnostic and with whom I have already had the joy of dialoguing for another text; and Gaël Giraud, a Jesuit. But this objective fact does not prevent them from carrying on an intense and constructive conversation that becomes the manifesto of a plausible future for our society and our planet itself, so threatened by the nefarious consequences of a destructive, colonialist and domineering approach to creation.

A believer and an agnostic speak and meet, albeit from different positions, on different aspects that our society must take on board in order for the world's tomorrow to be still possible: It seems to me something beautiful! And it is even more so because, in the unfolding of the discussion between the two interlocutors, there emerges clearly the conviction of the decisive importance of the word of Jesus, recorded in the Acts of the Apostles, not found in the Gospels: "There is more joy in giving than in receiving." Yes, because when the two interlocutors find in consumption driven to excess and in waste elevated to a system the evil of contemporary life, and identify in altruism and fraternity the true conditions for living together to be lasting and peaceful, they prove that the perspective of Jesus is fruitful and a place of life for all men and women. For those who have a horizon of faith and for those who do not. Human fraternity and social friendship, anthropological dimensions to which I dedicated my last encyclical, *Fratelli tutti*, must increasingly become the concrete and operational basis of our relationships, on a personal, community and political level.

The horizon of concern on which Petrini and Giraud focus their attention is the truly critical environmental situation in which we find ourselves, the child of that "economy that kills" that has caused the suffering cry of the Earth and the distressing and anguished cry of the world's poor. In the face of the news that reaches us daily—droughts, environmental disasters, forced migrations because of the climate—we cannot remain indifferent: We would be accomplices in the destruction of the beauty that God wanted to give us in the creation that surrounds us. All the more so because in this way that "very good" gift that the Creator forged from water and dust, man and woman, will perish. Let's face it: The reckless economic development to which we have given in is causing climatic imbalances that are weighing on the shoulders of the poorest, particularly in sub-Saharan Africa. How can we close the doors to those who are fleeing, and will flee, unsustainable environmental situations, the direct consequences of our immoderate consumerism?

I believe that this book is a precious gift, because it shows us a path and the concrete possibility of following it, at an individual, community and institutional level: The ecological transition can represent an area in which we all, as brothers and sisters, take care of the common house, betting on the fact that by consuming fewer things and developing more personal relationships we will enter the door of our happiness.

Vatican City, April 10, 2023

Francesco

INTRODUCTION

A book comes to life in various ways, sometimes unpredictably: a reading, an encounter, an influence. In this case, all of these things together: Carlo Petrini's in-depth experience, Gaël Giraud's intellectual radicalism and a bishop, Domenico Pompili, previously in Rieti and now in Verona, who connects them. A concrete opportunity presented itself: three days of intense reflection in a city, Rieti, the *umbilicus Italiae* (the navel of Italy), almost as though signifying that the words exchanged there by Petrini and Giraud, stimulated by questions, comments and inquiries, concern the very center of our existence: what to exist for, how to live.

The dialogue, which began in the summer of 2022 in the coolness of the medieval halls of Rieti's austere pontifical palace, was enriched by a visit to Amatrice, the emblematic town of the earthquake that struck Central Italy in 2016, an epiphanic phenomenon of the unhealthy relationship between humans and the environment that modernity has imposed on us. The critical reflection then continued through today's high-tech means, with video calls, emails, and virtual exchanges. As the reader will be able to verify, we have here the evidence that from two distant starting points, two people in dialogue can reach highly significant convergences, full of meaning and commitment for the dramatic present and the anxiety-ridden tomorrow. Yet it is a tomorrow that can transform, in ways both secular and Christian, into the colors of confident hope.

Gaël Giraud and Carlo Petrini are two intellectual activists with very different backgrounds, biographies and life trajectories. On the one hand, we have a learned Jesuit economist, mathematician and theologian, a professor at Georgetown University in Washington, DC. On the other we have an intellectual and professional gastronome, founder of the noted international Slow Food movement and the first University of Gastronomic Sciences. One a key international theorist of the ecological transition and a faithful interpreter of Pope Francis's *Laudato si'*; the other, the originator, in the name of good food and good living, of Terra Madre, the global network that brings together thousands of farmers, fishers and food producers from 150 countries around the world. Two key figures of our contemporary world, capable of coming together in a surprising way to outline a new, possible future, accessible to all. The extraordinary nature of this dialogue lies precisely in the ability to construct pragmatic visions and social practices of great impact, suggesting the design of a happy transition that each of us is called to build.

An exhortation attributed to Saint Francis of Assisi (although sources are not unanimous) resonated more than once during those July days spent in the shadow of Mount Terminillo, a stop along the Way of Saint Francis through the Holy Valley of Rieti: "Start by doing what's necessary; then do what's possible, and suddenly you are doing the impossible." The transition toward a fairer and more cohesive society is, in the end, not impossible, agree Giraud and Petrini. And it need not be based on sacrifice or deprivation. On the contrary, history is presenting us with the opportunity to fill our lives with justice, meaning and happiness. Failing to seize this opportunity would be improvident, especially in relation to ourselves in the dual sense of both individuals and members of an extended community. Above all, it would make us, our children and future generations, more unhappy.

This text is, at its core, a heartfelt call for communal action, for a collective mobilization of knowledge and practices that inevitably asks for more conscious and aware individual behaviors. There are two starting points: how we choose to feed ourselves and how we choose to behave within the economic-financial system. Giraud and Petrini teach us that these are two sides of the same coin, and perhaps also of the same cultural and social revolution that goes by the name of ecological transition.

Stefano Arduini

1. THE END OF A WORLD

CHAPTER 1

We Can't Go On Like This

Stefano Arduini This book arises out of an awareness that you both share: We are living in a world that is sliding towards ecological and social disaster with increasing speed. You both believe, therefore, that there can be no salvation other than truly embarking on the path of sustainability. On what grounds does this certainty rest?

Carlo Petrini Some elements need to be clarified immediately. We are on the threshold of a new historical period that will last not just a few years or decades. It will be a lengthy process leading to a change comparable to the Industrial Revolution. This is not about shifting from "bad" behaviors to "good" behaviors. That's not the point. Instead it's about changing the social model, embracing a new economic paradigm and radically altering the assumptions and reasons behind our way of life.

The Industrial Revolution spanned over a century, from the second half of the 18th century to the end of the 19th century. Its foundations were a process of economic evolution and industrialization that transformed agricultural, artisanal and commercial contexts into systems characterized by the widespread use of machines powered by mechanical energy and a reliance on fossil fuels. All of this was facilitated by a strong component of technological innovation. Humanity enjoyed significant improvements in well-being, health and education due to this epochal change. However, in order to transform and produce, the Industrial Revolution was built on a myth: that Earth's resources are infinite. A myth that has now shattered, because we have reached a point of no return. We must open our eyes. We are facing a dramatic and irreversible situation. The ecological transition towards a sustainable model is the only possible path forward. The task may be challenging, but an aware civil society can achieve it.

Gaël Giraud The most important thing to understand is that the alternative is not business as usual; the alternative is global disaster. It may not be evident in public discourse, but already within scientific circles there is talk of the potential extinction of humanity in the next century.

Let's suppose we continue to behave as we have done so far. From eastern Siberia and the Arctic Ocean, we have confirmation that the large-scale release of methane from permafrost, the soil that should be permanently frozen, has begun. If all of this methane is released, we might have a temperature increase of 6 to 7°C by the end of this century, and by the end of the next century, the rise will be in double digits.[1] That means that we are moving perilously close to the extinction of humanity. An example? As the temperature rises, the availability of water decreases. Already now it is clear to everyone that water scarcity is a serious problem for Italy, for example. According to the WEI (World Engagement Institute), without intervention, by 2040 there will be a 20% decrease in water availability worldwide compared to today.[2] Meanwhile, the frequency and severity of extreme events such as typhoons, floods and desertification will increase. Faced with this likely disaster, we cannot close our eyes. By now, most scientists are aware that continuing along the business-as-usual path is suicide. This point should be clear to everyone: It's no longer a matter of being optimistic or pessimistic, but simply scientific realism.

The ecological transition is, therefore, a duty, unless one wants to fall victim to an obsession with disaster. To be honest, this fascination still exists to some extent in certain circles and is very dangerous.

C.P. At this point, it's important to clarify one of the key terms in our discussion: sustainability. Sustainable shares the same root as "sustain," the name for a piano's right pedal, which prolongs the resonance of a note. The French translate it as *durable*. That term embodies the idea that the actions we undertake should have long-lasting, durable results. In contrast, the blind pursuit of a capitalist logic, embedded in the globalization race of the last 70 years, has, among its numerous effects, led to the formation of economic and production models that are unsustainable due to their short duration. However, those who, as our French counterparts teach us, translate the word "sustainability"—which by now is in all too common usage—as "durability" can easily recognize, with the naked eye, how dramatically our society has been impoverished in recent decades. With industrialization and the indiscriminate use of natural resources, we have triggered, in a domino effect, a climatic-environmental crisis and a rampant loss of biodiversity unparalleled in history, undermining our very survival. As Gaël Giraud said, we are immersed in a suicidal socio-economic dynamic. And we must break free from it.

So we must embark on a new path. But changes of this magnitude do not happen with a magic wand; they need a period of growth and awareness, hence the involvement of communities. How long will this phase last? The entire process? I said it before: We don't know, no one does. It will last as long as it needs to last. What is certain is that the journey has already begun; a population has set itself in motion, driven primarily by the protests of many young people asserting their right to live in a healthy environment. This gives me great confidence in the future.

G.G. I'd like to give a personal example. For some time, I've been carrying a very concrete anxiety within me. The idea that by 2050, Mont Blanc will green up considerably, haunts me.[3] Similarly, I cannot bear the thought of the disappearance of bees and fish, and the suffering caused by the lack of access to drinkable water mentioned earlier. The disaster of this model is right in front of our eyes.

Let's consider another point of view, the demographic one. From now until 2050, sub-Saharan Africa will account for 57% of global demographic growth. Consequently, about 23% of the world's population will be sub-Saharan, compared to the current 15% and 10% in 1990. The share of the global population in the European Union, which is currently around 6%, will decrease to 4% in the next three decades. This means that in 30 years the African population will grow from 1.4 billion to 2.4 billion, according to the UN median scenario.[4] High fertility rates and improved life expectancy are supporting an extraordinary pace of demographic growth that will result in a drastic increase in the demand for social services in countries with high levels of poverty and economic inequality. Moreover, rising temperatures and humidity levels from ongoing global warming will make some parts of the world uninhabitable, like sub-Saharan Africa, Amazonia, Central America, the southeastern coast of the United States, parts of China and India and a significant portion of Southeast Asia. All of these regions are already densely populated and are seeing rising demographic rates.

We need to reflect on the consequences of mass migration, which could lead hundreds of thousands of people to migrate to Europe. There is no possibility that the border defense policies of right-wing political parties can succeed, beyond propaganda. Instead we need a strategy, and as Carlo Petrini believes, ecological and social transition is today the only possible path that can spare us the anxiety of facing enormous problems without any tools to tackle them. It is along this path that we should proceed. Or perhaps, as Carlo suggested, it's a path along which we have already slowly started.

C.P. It is right to look at migrations to realize how disturbed politics and society are by a phenomenon they cannot comprehend or govern. But, if we broaden our horizons, we can see that this dynamic does not concern only humans or other animals. Wine offers a case in point. The increase in average annual temperatures by just 1°C shifts the optimal habitat for grapevine cultivation by 150 meters [492 feet] in altitude and 150 kilometers [93 miles] in latitude. This means that in Europe, vineyards are moving increasingly northward and even pushing into England. In my homeland of southern Piedmont, for example, they are moving up into the Alta Langa: Where once there was only pasture and nothing more, now there are vineyards everywhere. But all in all, at our latitudes, there is neither loss nor gain. The budget is balanced: We will probably adapt to cultivating something we once considered exotic. In sub-Saharan Africa, however, we must prepare for a disaster on all fronts. And it would be illusory, as well as profoundly unfair, to think that these will be problems only for Africans. Pastoral communities in Tanzania, Uganda and Kenya have no prospects for long-term survival there. By 2050, we will have over 150 million climate migrants.[5] And some among us think that the way to address a problem of this magnitude is not with policies against global warming but with naval blockades. If millions of people are unable to survive in their home countries, no blockade will hold. It is not clear what we are waiting for to open our eyes.

S.A. Especially considering that there are now numerous international reports stating that water crises are one of the main causes of migrations, like dossiers from NGOs such as GVC, Oxfam and Save the Children or the World Economic Forum's 2016 *Global Risk Report*, which highlighted the water crisis as a decisive factor in migratory movements. Since 2009, it is estimated that one person every second has had to leave their home due to environmental catastrophe.

G.G. On paper, the numbers alone should be enough to determine the way forward. We have already mentioned many, but here are some additional figures that I believe are significant, taken from *Water Grabbing*, a compelling book written in 2018 by Emanuele Bompan and Marirosa Iannelli. Approximately 1.6 billion people live in countries with limited water availability, and predictions suggest that this figure could double in two decades. According to the International Food Policy Research Institute, between 650 million and 1.3 billion people live in basins with high levels of nitrogen, phosphorus and arsenic.[6] The "economistic" paradox is that, according to the World Bank, water-related migrations will result in significant GDP losses estimated at between 7% (East Asia and Central Africa) and 14% (Middle East). A dramatic decline of about 11% is expected in the Sahel and Central Asia, while a 2% decrease is hypothesized in Southeast Asia.[7] What's the result? Where water is scarce, poverty increases and consequently people seek to migrate.

CHAPTER 2

Too Much Food and Too Much Hunger

S.A. Your analyses and studies put two sectors in particular on trial: the food industry and financial markets. Let's start with the first . . .

C.P. I am convinced that the main culprit for the environmental mess is the food system, along the entire production chain: agriculture, livestock farming, processing, distribution, consumption. People usually tend to believe that mobility (cars, trains, planes, etc.) is the main contributor to global pollution. That's not the case. Worldwide, mobility accounts for 17%, while the food system accounts for 34%.[1] Twice as much. Globally, we produce food for 12 billion human beings. However, Earth's population is currently 8 billion. The result? 33% of the food is wasted.[2] 33% means millions of tons of food being produced for no reason, stressing acres and acres of fertile land and using billions of liters of water. It's madness with no rhyme or reason. There are moral responsibilities that are as obvious as they are enormous. Of course, these responsibilities weigh differently in the global North and the global South. Here where we are, waste is functional to an economic model that considers food a low-value product. We overproduce so that supply always exceeds demand and prices remain low. Essentially, waste is accepted because it is good for the economy. In the global South, on the other hand, waste is due to inadequate infrastructure, because there is a lack of transport and, in hot regions, appropriate tools for food preservation, as we have seen dramatically with vaccines.

Another example: the waste of water. We have already talked about its scarcity. Now I want to focus on a different aspect. The next wars will break out over the hoarding of water resources, not the conquest of land. Today, saying "take one less shower" or "turn off the tap when you brush your teeth" is correct because we are talking about good practices, but it's not enough. Personal water consumption accounts for 9% of the total, industrial consumption is 22% and agricultural consumption is 69%.[3] A synergy between personal choices and economic, political and social choices is necessary to truly make a difference.

Returning to the point, the food supply chain is where the decisive game is played. One-third of water consumption is dedicated to the intensive farming of chickens, cattle and other animals that end up on our tables. More water is used for livestock farming than for growing vegetables and fruit. It is evident that, from the perspective of public well-being, none of this makes any sense.

This is a criminal and failed system. I'll add a couple of data points provided by the United Nations. There are nearly 900 million malnourished people worldwide who do not have food security, and 200 million people literally suffering from hunger.[4] In the same world, however, there are 1.7 billion people suffering from overnutrition, obesity (13% of the global population) and cardiovascular diseases:[5] It's a snapshot of the madness in which we are immersed. According to FAO data, by 2030, the percentage of people suffering from hunger will be the same as in 2015: 8%.[6] This means that despite the speeches, the many words spoken and the many promises made, nothing will have changed in 15 years. Let's look at the facts. On the one hand, there are those who die of hunger, on the other, those who live with diseases caused by overeating or a poor diet. We must painfully acknowledge that malnutrition is not due to a lack of food, but to poverty. Food today is not for everyone; it is not clean, considering that a third of greenhouse gas emissions are linked to the food supply chain; often, it is not even particularly good.

Let me make a brief digression on highly processed foods, those rich in salt, sugar and fat. These foods are extremely harmful and create addiction perhaps even faster than smoking, alcohol or other drugs. A study has demonstrated that cigarette smoke takes ten seconds to activate brain receptors, while sugar takes less than a second.[7] The faster a substance acts on the brain, the higher the likelihood of an impulsive and compulsive reaction. Behind hyper-processed foods, there are also deep dynamics that perpetuate injustices. Often, those most dependent on junk food are also the most vulnerable, with a low level of education and without the time or money to spend on preparing or buying healthy, high-quality food.

But I remain convinced that 8 billion people can live and feed themselves sustainably. I mean sustainable in the most authentic sense I described earlier: using resources in such a way that they continue to be available in the future. Eating in a sustainable and durable way means ensuring that what we consume today can continue to be enjoyed by our children, starting from the soil, the origin of all food. For agriculture to be sustainable, for example, it must abandon pesticides: poisons that kill fertility, in addition to being harmful to our health. The dominant model of the Green Revolution (large-scale production based on extensive use of machinery and a reliance on pesticides and fertilizers), which began with the end of World War II, has reached the end of the line. Those who claim that chemical pesticides are necessary for food security are spreading misleading information. Even though per-capita food availability worldwide has increased in the last 40 years, we have impoverished the soil without eradicating the problem of hunger. That's not me speaking, but leading experts from the United Nations. An excess of synthetic chemicals has doped agriculture, weakened the ecosystems that support it and made farmers dependent on large multinational corporations. All of this has taken place in a world where more than a billion people still suffer from hunger.[8] Leading agriculture into the era of ecological transition is a moral duty, and fortunately, it is also possible.

In the world, there are already many virtuous agricultural enterprises: consider that over half of the population is fed by 500 million small-scale producers, family businesses or cooperatives.[9] This immensely valuable fabric should be safeguarded, protected, defended, supported and promoted, but is increasingly strangled within a system that favors multinational corporations, agribusiness and the big players in applied food chemistry, the ones who not only own patents and hybrid seeds but also tend to receive a large portion of the funds allocated at the international level.

To feed eight billion people, the path is as clear as it is revolutionary: stop chasing economic gain and start defending food production and the land on which it depends. Food must be a right, not a good to be traded on the stock exchange; a universal priority, not a commodity for enriching the few at the expense of others, the health of the planet and the future of humanity.

CHAPTER 3

A Planet of Plastic

S.A. One material that is central to our reasoning: plastic. Of course, there is more to plastic than just negative aspects. As marine biologist Silvio Greco explains in his book *La plastica nel piatto* ("Plastic on the Plate"), not only does plastic packaging contribute to reducing food waste, when used in wind energy installations or solar panels, it plays an important role in the development of renewable energy. However, "the almost non-existent plans for proper disposal have caused an increase of plastic pollution in the marine environment, leading now to the explosion of a real global emergency. In 2015, the annual global production of plastic resins reached 322 million tons, an astonishing acceleration considering that in 1980 annual production was just over 50 million tons."[1]

He continues: "In 2016, according to World Bank data, global production of urban solid waste (domestic and commercial) reached 2 billion tons and this is expected to double by 2025. Half will consist of non-biodegradable materials (plastic and metals). The reasons for this overproduction of waste lie on the one hand in their disastrous handling on land, and on the other, in an equally thoughtless daily use of plastic, which is now the major pollutant in the planet's seas."[2]

C.P. I'm familiar with Greco's work. He also tells us that "it is estimated that at least 5.55 trillion pieces of plastic are floating in the world's seas and oceans. If this trend is not regulated, changed and reversed, by 2050 the total amount of plastic dumped in landfills and dispersed in the environment could amount to 12 billion tons. And the problem will be inherited by future generations. On average, between 5 and 13 million tons of plastic end up in the sea every year, as we have seen. Based on the estimated weight of the world's fish stocks, at this rate, by 2050, there will be more plastic in the oceans than fish."[3]

All of these facts and figures, one after the other, speak for themselves. Little is written about it, but one of the major issues that needs to be addressed is the health consequences of plastic pollution, which has an immense impact on the oceans and has reduced the Mediterranean to a pitiful state. By eating fish, mollusks and crustaceans and even simply by drinking water, plastic has entered our food chain. We need to be clear: We are eating microplastics, and we are habituating our children to eat microplastics. An important study commissioned by the WWF from the University of Newcastle has shown that through microplastics we ingest the equivalent weight of a credit card every week.[4] We are eating plastic, and where does the fault lie? Mostly in single-use plastic, used for food packaging:

Putting three slices of ham in a plastic tray, which then ends up in the trash, is not only completely senseless but also dangerous for our health. Yet we continue to produce and buy it.

S.A. Let's try to provide some additional elements for reflection. In 2021, another book was published that can help us better define the phenomenon, entitled in Italian *C'è vita senza la plastica. Perché farne a meno prima di morire soffocati* ("There Is Life Without Plastic: Why do without it before suffocating to death"). It was written by journalist Hélène Seingier together with Nathalie Gontard, one of the world's most renowned scientists in the field of plastic and plastic recycling. According to her calculations, every Westerner consumes about 100 kilos [220 pounds] of plastic per year; since it began to be manufactured on an industrial scale in the 1950s, 9 billion tons have been produced. If the current trend continues, we will reach 30 billion by 2050:[5] dizzying numbers for a material whose implications in terms of pollution we still do not fully know. The plastic produced in the past, Nathalie Gontard claims, has been worn down and is now nanometric in size: Will we have a world infested with microplastic particles for years to come? "In the first decades of plastic bulimia, no one asks questions, no one looks far ahead," writes the scientist. "Focused only on benefiting from the successes of his intelligence in exploiting the Earth's resources, man interprets happiness in a material sense. His view is limited by the blinders of economic growth: He cannot foresee the consequences of the massive use of plastic across the entire planet."[6]

Already today, the numbers and examples are terrifying: Even in the Mariana Trench, at over 11,000 meters [36,000 feet] below the surface of the sea, 100% of crustaceans show traces of plastic material; a "new continent" of plastic fragments three times the size of France was identified in the Pacific Ocean in 1997; in Australia, certain soils contain up to 6.7% of PVC microplastic particles while 83% of rivers in the United States are contaminated by plastic. What can be done to escape the tyranny of plastic? Gontard has no doubts: invest in biodegradable materials, given that plastic is the first material outside of nature produced by humans: "It cannot find its place in the great cycle of nature—or, at least, not before persisting for many centuries in the environment, threatening to cause enormous damage in the meantime."[7]

CHAPTER 4

Economics of Suffering

S.A. After lifting the veil on the fatal flaws of the food production chain, let's move on to the second accused: the current unsustainable financial model.

G.G. We need to define the playing field, in other words to understand what is neoliberalism, the ideology that currently inspires much of the world and all financial markets. Neoliberalism is attempting and—as far as we can say to date—succeeding in destroying the legacy of the Enlightenment. It did this when it introduced a new imperial code, namely the financial market, which, through the weapon of public debt, imposes its policies on states, especially the most compromised, like Italy.

Let's be clear about this. Many of the same adjectives used for God have in fact been transferred to the financial market. Some economists define the market as "omnipotent" or "omniscient," sometimes "benevolent," but in any case, they speak of it as if it were the demon Moloch from the Old Testament. A kind of pagan religion has been reintroduced, holding banks and business up as untouchable deities. In this way, neoliberalism destroys another pillar of modernity because, in practice, it undermines equality before the law. Scandalous tax evasion practices occur in all countries these days and tax havens exist even in the Euro area. Ireland and Luxembourg, for example, simply allow big companies to pay much less tax than they would in their countries of origin.

This is exactly how a sense of equality and the rule of law is lost. Consider the 17 Sustainable Development Goals (SDGs) established during the United Nations Summit in New York in 2015. No reference to democracy and human rights is made in any of the 17 items. It's as if the United Nations thought development could take place without democracy. I believe this is an eloquent example. I reiterate, the neoliberal spirit has demolished the second pillar of modernity: equality before the law, one of the foundations of human rights.

C.P. The other side of neoliberalism is consumerism. An ideology that depletes resources, creates waste and paradoxically does not satisfy people's needs. This is the other decisive point. In the globalized world, it has reached its peak: We ourselves are now consumer products, and therefore we can be, we deserve to be, plundered, wasted like a mass-produced consumer product, passively mistreated without ever achieving a true state of well-being.

In the food sector, this logic has imposed an insane principle: Food must cost as little as possible. When the price increases, even slightly, public opinion, newspapers and politics are ready to take to the barricades. If cell phone tariffs or banking fees increase, on the other hand, we remain silent. This happens because we can no longer see food's material, cultural and spiritual significance: The neoliberal system establishes its value solely through price. The result? Food is no longer produced to eat, but to sell. Food has become a commodity like any other. But instead, we must understand that food contains within itself a universe of values that go far beyond its price. Making a jar of jam at home certainly costs more than buying it ready-made, but it gives us much more in terms of personal gratification, service to the community and environmental protection, in other words in terms of well-being.

In the name of the "money god," we have built a monster: Food, stripped of its ancestral meaning, ends up being something hostile. Without its cultural, social and environmental values, without the care and pride it deserves, it transforms from a resource into an evil creature that devastates the countryside, creates injustice and spreads pain. This consumerist and neoliberal food can be easily wasted, but it leaves us alone in our uncertainty and unease. Giraud spoke before of anxiety about the scale of the disaster we are building. We become restless because we have fallen prey to false needs induced by advertising and marketing, the consumerist monster's powerful weapons. A child, but also a teenager or an adult, who spends hours in front of a screen—whether a television or any other device—is subjected to a media bombardment in which food appears to be something it is not. Like the case of the cow coming into the kitchen to give us milk in the morning, or the grandmother pushing her family out the door because she claims not to have done the shopping and sitting down alone to enjoy a ready-made dish. Yes, the grandmother, who in our social and culinary tradition is the guardian of the knowledge of flavors and taste; she is the one who is supposed to imprint us with our gastronomic education, transmitting the joys of food. Advertising that convinces us of needs we didn't even know we had offers the clearest evidence that we are no longer masters of our food. Nor our future.

S.A. How can we get out of this trap?

C.P. In the second half of the 20th century, Western society chose to prioritize a hyper-individualistic model. This led to us assessing the richness of our lives through sterile indicators such as GDP, net worth and profitability. Not surprisingly, these are quantitative parameters on which no limit is imposed. On the contrary, the dominant model wants to instill the belief that the more we push toward unlimited growth, the more widespread well-being can be enjoyed. But today we are living in the proof that this approach, essentially based on competition, not only failed to meet its objective but has revealed side effects that have negatively impacted social relationships and relationships with the ecosystems in which humans live.

Mending our interaction with natural resources—which are not infinite and are primarily common goods available to all human beings—and rebuilding the foundations of our society on relational goods such as compassion, altruism and fraternity represents the greatest challenge of our time. Only in this way can we again understand the true meaning of the concept of well-being, which inherently relies on the qualitative aspects of individuals' lives. In other words, once physiological needs are satisfied, we should aspire to "be more" rather than "have more."

All of this, in my view, can only happen if accompanied by the development of a proper sense of limits, strongly correlated with the maturation of human relationships. Without empirical knowledge of our own limits, boundaries and the contexts in which we immerse ourselves, it is not possible to create healthy and fruitful interpersonal relationships. At the same time, it is precisely when humans find themselves in social environments characterized by a high level of cooperation that the demarcation lines of cooperativism are created. These are not weaknesses but rather tools to enhance what we can define as common goods.

Current economics must make room for a model based on consolidation rather than growth. This is possible by embracing the concept of limits, whether they relate to production, consumption or cost, and by giving centrality to common and relational goods. With all due respect to those engaged in greenwashing, riding the sustainability wave, it is time to acknowledge that only practices implemented by those who learn to live in equitable and shared prosperity with all other humans, within Earth's physical and biological limits, are truly sustainable. This is the only real trend we can afford for the future.

G.G. The political program for a European modernity 2.0, the project for the society of tomorrow, involves overcoming neoliberalism and detaching from it, in other words an ecological transition that "cuts off the king's head."[1] Specifically, it means seriously regulating finance, restoring a rule of law where equality and respect for human rights prevail and ultimately replacing private property with common goods. In Italy, reflections on the concept of the common good have developed mainly under the influence of legal scholar Stefano Rodotà, who played a fundamental role in having water seen as a common good within constitutional principles. I remember how in 2011, in Naples, a very strong civic movement rose up for the municipalization of drinking water and against its privatization. Similar movements developed in Bologna, Rome, Milan, Bergamo and other cities.

We can also interpret the experience of Slow Food as an action aimed at asserting that food is also a common good, not a private one. Certainly, we can interpret the *Laudato si'* encyclical as an exhortation to pay attention to and care for the environment, a precious common good. Finally, the Covid-19 pandemic has shown that health is also a universal common good, which was evident when the disease that affected someone in China was the same that then impacted all the inhabitants of the world.

S.A. A cultural change of that dimension presupposes a mature education system able to support it. Are we really ready?

C.P. When it comes to gastronomy and food education, schools have enormous limitations. Food and gastronomy are not part of the curriculum, while teaching nutrition often brings about the opposite of the desired effect. School textbooks fall short of their intended function: They are often repetitive, boring and abstract. How can one engage a student with nutritional tables or videos on what is good and what is bad, especially when unhealthy foods are presented as the most attractive? These lessons become lectures filled with information about how our bodies work, distancing students from the pleasure and richness of food and gastronomy. In short, it's counterproductive! Teaching children where food really comes from, letting them experiment by handling, cooking and tasting, is the only way to educate them about pleasure and taste. It's essential to value and guide their sensory experiences so that they become familiar with the food culture of their region. In this way, they are equipped to discern differences and, consequently, make informed choices when purchasing.

CHAPTER 5

Toxic Finance: The Role of the Banks

S.A. We'll return to sustainability as an educational challenge later, but for now I would like to concentrate on the dynamics of the banking world, a subject on which you, Professor Giraud, have reflected and written extensively . . .

G.G. Neoclassical economic science emerged around 1870, yet in a century and a half, it has in no way modified its postulates, some of which are completely false. Everyone knows this, but still, in 2022, the Nobel Prize in Economics was awarded to three economists (Ben S. Bernanke, chair of the Federal Reserve, the central bank of the United States, from 2006 to 2014; Douglas W. Diamond and Philip H. Dybvig) whose theory of money and the banking system is not accurate.

One postulate of neoclassical economics is the neutrality of money. According to this position, an increase in the volume of money in circulation has no impact on the real economy except for a proportional increase in prices, across the board. In other words, if inflation increases and the money supply doubles, prices will also double. A simple multiplication by two. In this context, the relationship between the relative prices of a bottle of water and an English language course—just as an example—remains unchanged as inflation rises. Unfortunately, this is not the case, and it can be empirically verified; there's no need for extensive studies. The impact of money affects economies and, consequently, people's lives. Nevertheless, in all models used by the International Monetary Fund, the European Central Bank, the Federal Reserve and the World Bank, money is considered neutral.

Then there's another mystery. Within these models, it is taken for granted that banks cannot create money. Money is considered something magical, without origin. The truth is that banks create money every day; it's their job. When a bank extends credit, it is creating money for you. This is the most normal thing in the world, but it is not taken into account by neoliberal models. And here we come to the point.

If we want to finance the ecological transition, we need a lot of money to support investments in green infrastructure. We estimate that approximately $90 trillion is needed globally.[1] This is roughly equivalent to the global GDP. And we need it by 2035. A dream? Not at all. If we need $90 trillion, then we need to understand how to achieve this goal and which institutions can create money to fund the ecological transition. But if, as the neoliberal paradigm would have it, we can't even discuss it, then it's obvious that this necessary epochal transformation cannot happen. And this is where we come to banks, currently adversaries of the ecological transition because they are trapped in a false and perverse economic and social logic.

S.A. So you are saying that the banks are not "evil" per se, but are instead part of a flawed system?

G.G. In 2022, I published a report on the top eleven banks in the Eurozone, two of which are Italian. From that study, it is evident that their total investments in fossil fuels add up to €530 billion, equivalent to approximately 95% of the sum of the market capitalization of each of the institutions taken into consideration.[2] The consequence is that if we wake up tomorrow morning and ban the consumption of coal, oil and gas, all these major financial entities would go into default. Do you know what that means? Simply put, the business model on which banks operate is entirely incompatible with the ecological reconstruction of the European continent. My analyses are not isolated. For instance, according to Oxfam, an international confederation of nonprofit organizations advocating for the reduction of global poverty, out of every €10 of credit allocated to energy projects by French banks, as much as €7 are directed towards projects based on fossil fuels (2019 data). For these reasons, if we do not change the model beyond some window-dressing and greenwashing, we will not make any progress. The question we must now ask ourselves is: How do we change course?

I'll try to use an analogy to explain: Our banks are like a smoker who has developed metastatic lung cancer. How to intervene? I argue for using the ECB, the European Central Bank, as a surgeon to remove the diseased cells. First the patient must be convinced to quit smoking. But we cannot impose a ban on cigarettes without removing the metastases. In other words, we cannot ask credit institutions to operate only in green markets "from now on" without relieving them of their previous financial exposure to fossil fuels. If we did that, banks would undoubtedly react negatively, and we would have failed in our attempt to have them as allies in the path towards ecological transition. The numbers are staggering, and we must always keep in mind the scale of this exposure to fossil products, which by 2035 will reach $90 trillion globally.

S.A. You mentioned greenwashing. How common is this phenomenon?

G.G. There is a very specific strategy behind greenwashing. To understand it better, we need to go back to the financial crises of 2008–2009. The banks, finding themselves called to account, had to give the impression that everything had changed. That's when, as if struck by a revelation on the road to Damascus, they "discovered" ecology. But it could only be a deception. Because, as we have seen, if they truly continued on this path without being thrown the lifebelt we talked about, they would be bringing about their own demise. The result is that today, the world of credit sees ecological transition as a formidable marketing and branding tool, without this communication and media exposure having any real fundamental impact.

S.A. Another key point in your overview of the crisis of neoliberalism is related to the actual level of employment. Can you explain this further?

G.G. Neoliberals do not admit that there is involuntary unemployment in an economy. According to them, the meeting between employees and employers functions like a transparent, immediate and universal auction, where the wage level marks the equilibrium point between supply and demand. People who remain unemployed are those who have not accepted the equilibrium point. In other words, they are unemployed by choice and therefore classified as voluntary unemployed.

The problem is that this model does not reflect reality because it is entirely evident that even if we accepted the metaphor of the auction, it is certainly not universal and transparent. It is equally apparent that only a minimal number of the unemployed could afford to reject a job. Moreover, besides being unreal, the model relies on figures that do not correspond to the actual numbers of those without a job. As we write this book, the official unemployment rate in my country, France, is 7.3%, and in Italy, it is 7.8%. What is not mentioned is that this percentage considers as employed even those who, for example, are paid for one hour's work per week. But part-time jobs almost always result not from the worker's choice but from the structure of the labor market. In France, and the same goes for Italy, there are many women who do not work full-time simply because they have not found full-time employment. If the statistics correctly accounted for this "non-working" time, the employment rates would be reduced to 70% in France, Italy and the United States, and 75% in Germany.[3] This means that unemployment remains a severe ailment of the capitalist economy, completely ignored by the classical and neoclassical models with which economists work. Becoming aware of this would completely change the analytical framework and, consequently, labor policies.

S.A. With what concrete consequences?

G.G. All of the reasoning we have developed is completely ignored by the majority of economists. The consequence is that none of them will seek a solution for unemployment at 30%, simply because, for them, the problem does not exist, given that the official rates are much lower.

For them, the problem lies in public spending. And here, we touch on another significant hoax imposed by the neoliberal narrative, namely that public debt represents a mortal sin.

S.A. Let's look at Italy, which has over €2.5 trillion of public debt, equal to 145% of its GDP.[4] Can we really not be alarmed by these numbers?

G.G. I don't think it's a problem. It's not a problem because the European Central Bank can create money and therefore save any country. You just need to want to do it.

Before the creation of the Eurozone, there was an Italian central bank in Rome that had the power to create currency for the national treasury. The same was true in France and other countries. After the Second World War, an entire generation rebuilt Europe using inflation, without letting public debt spiral out of control because each central bank could create the necessary currency for the national treasury with an interest rate equal to zero. And when the national treasury had to repay the central bank, it was very easy to return the money with a new loan from the same bank. At some point, in the 1970s, some of the European elites decided that this mechanism had to end because it gave the state too much power and their goal was to limit state interference in the markets as much as possible. How to achieve this goal? By preventing the state from borrowing resources from central banks at zero interest, so that it had no choice but to rely on financial markets, naturally recognizing their positive interest rates.

What is almost never said is that this new paradigm has increased the debt of all European countries: Italy, France, Germany and so on. In France, for example, today 40% of the public debt is determined by the interest rates demanded by the market.[5] If we had continued to apply the practices of the old post-war system, in France we would only have 60% of the public debt that we see now. I am sure the same can be said for Italy. This means that the increase in public debt is, for the most part, only the consequence of the creation of the Eurozone and the misguided decision that the European Central Bank does not have the right to create money for a state that needs it.

S.A. So it's all the fault of Europe? And, most importantly, what should we do to break free from the grip you've described?

G.G. There aren't many alternatives. We must be fully aware that it is perfectly possible to erase public debt through the European Central Bank (ECB). I say this because it has already been done in 2009 after the subprime mortgage crisis of 2007 and 2008. In that case, the ECB, through quantitative easing, created money to save the banks that were excessively exposed in the mortgage market and were therefore in very worrying situations. Without going into technicalities, the consequence is that today there are billions of public debt on the ECB's balance sheet. It's madness because the states, in addition to being shareholders of the European Central Bank, are also debtors. I repeat: The shareholders are debtors of the ECB. It would, therefore, be perfectly possible to erase these debts from the ECB's balance sheet overnight. These are political choices. There is no reason why an international European summit couldn't be organized, with German Chancellor Olaf Scholz, French President Emmanuel Macron, Italian Prime Minister Giorgia Meloni and all the other European leaders, to agree on the immediate cancellation of public debt and to decide to use the "freed" resources to finance the European Green New Deal. When I proposed this in France, some economists argued that it would be a disaster for the euro because the ECB would lose its credibility, and the consequence would be a complete failure of our currency. The paradox is that today the euro remains weaker than ever precisely because international investors think that Europe is lagging behind in the ecological transition that the US under Joe Biden and China are planning, if in their own way. The truth is that if we don't seize this opportunity, then yes, we will accelerate the decline of the euro.

S.A. Is China really ahead of us in the ecological transition?

G.G. It's a paradox because globally, China is the country where most green investments are concentrated, but at the same time, it's the leader for emissions, with nearly 10 billion tons of CO_2 released into the air, accounting for 28% of global emissions. However, I reiterate, when we look at the percentage of GDP invested in green policies, China is ahead of us.[6]

S.A. Previously, Carlo Petrini criticized Gross Domestic Product (GDP). He is not alone. Many now consider it an unreliable indicator. What is your position as an economist?

G.G. GDP is a completely flawed indicator. Let me give a very simple example. According to orthodox economists, a car accident, an event that can have tragic consequences, would have a positive impact on the GDP curve because it corresponds to an increase in consumption and therefore production: think about medical transport, hospital expenses, rehabilitation and medications as well as the costs of repairing the road and the vehicles involved in the accident. Just this example illustrates how an increase in GDP, by itself, does not mean an increase in people's well-being. On the contrary, in the case of our accident, it would mean exactly the opposite.

We need improvements in fields like healthcare or education, which don't necessarily have a direct relationship with GDP. Its increase in the last 20 years has been primarily a consequence of the financial bubble. And the financial bubble has not helped people's well-being at all, except for that very small fraction of the privileged who earn immense sums on the global markets. This doesn't negate the fact that financial markets remain entirely inefficient because they are based on flawed calculations.

Let's stay with GDP. The way to calculate this measure of size has no mathematical basis. It is therefore reduced to a trap to drain economic value for the benefit of financial investors. There are three ways to calculate it: the spending method, the value-added method and the income method. Each is different and they are not equivalent. However, economists need them not to contradict each other. What to do? Very simple: They adjust the numbers to make the math work out. I call it "the dirty kitchen of statistics."

This alone would be enough to undermine the credibility of our economy's main indicator, but there's more: Not only is nominal GDP entirely unreliable, but if we move from nominal to real GDP, we discover that the latter has no meaning. Real GDP should be the difference between nominal GDP and the inflation rate. If prices are stable, there is no inflation, and therefore nominal and real GDP overlap. In reality, however, prices fluctuate. Let's say there is an inflation rate of 2%, and at the same time, nominal GDP always increases by 2%: This should mean that real GDP remains constant and unchanged. To show that this is indeed the case, however, one would have to verify each year the relative prices of different consumer goods, for example, between a can of tomatoes and the cost of an hour of English language class. If economists did this, we would have a credible real GDP. Given the economic and time costs, however, it is impossible to constantly benchmark relative prices, especially in situations like the recent significant increase in inflation due to an energy shock that led to a significant rise in prices for energy but not other products or services. The result is that we have in hand a completely hypothetical indicator that means nothing at an economic level. Nevertheless, real GDP still constitutes the main objective of all public policies in Europe, in France, Italy, Germany and so on. In other words, we are organizing our societies based on a false indicator, which does not take the principles of sustainability or social and environmental impact into account in any way whatsoever.

But the worst thing is that it's not true there are no alternatives to GDP. There are plenty of them. We could list a hundred. For example? Canadian economists Lars Osberg and Andrew Sharpe have developed an economic well-being indicator based on four axes: levels of consumption integrated with unpaid domestic work and volunteering, the ability to take care of the environment and two other factors that measure how the well-being of each person decreases when economic insecurity and poverty increase.[7]

Then there's the ecological footprint, an indicator from the green world that aims to quantify the impact on the environment of every human activity. Just to understand: a flight from Paris to New York, which would delight GDP supporters, is equivalent to the average ecological footprint of one person on the planet in one year. According to this indicator, if 8 billion people adopted the average lifestyle of North Americans, we would need five Earths to survive. Another alternative indicator is the BIP 40, which stands for the barometer of inequalities and poverty. The model was developed in France by a network of researchers, activists and trade unionists and includes sixty variables and six dimensions of inequality: housing, health, education/training, justice, work and income. For now, I'll limit myself to these three examples, but I could provide many more.

S.A. What's stopping states from replacing GDP with a more reliable indicator?

G.G. With this question, we get to the heart of the problem. Let's say that starting tomorrow, Italy does what Bhutan, a small Buddhist kingdom on the southeastern slopes of the Himalayas, does, which is to not publish GDP but instead provide a very interesting happiness indicator. If Italy did that, it would step out of every politician's favorite game: comparing their GDP with that of other countries (or with that calculated by their internal competitors) to demonstrate that they have been a good administrator. Then, of course, a gigantic issue would open up in regards to the economic and financial relationships between states and between states and private entities that would be unmanageable unless all countries simultaneously decided to shelve GDP. In which case an alternative would already be at hand: the Human Development Index (HDI). This macroeconomic development indicator represents a different way of evaluating the well-being of a nation, taking into account factors beyond GDP per capita, including life expectancy at birth, average calories available per person, access to clean water, literacy and schooling rates, health services and the degree of political freedom. It is well known in the United States and, in my opinion, would be the easiest alternative—not the best possible indicator (that's why I didn't mention it earlier), but still much better than GDP.

S.A. You are an economist but you've also studied mathematics and worked in finance. In your book *La rivoluzione dolce della transizione ecologica* ("The Sweet Revolution of the Ecological Transition"), you wrote that banks move their mega-computers to basements to get physically closer to the places where market orders are recorded, hoping to gain a few seconds over competitors.[8] Is this really enough to make a difference in digitalized financial markets?

G.G. It's not just a few seconds but rather milliseconds or microseconds. This mechanism has a specific name: high-frequency trading, which now accounts for half of the trading in US financial markets and 30 to 40% in Europe. It's a market where there are no longer humans deciding whether today it's worth buying the stocks of a bank or an automotive company. Computers are in control. They make decisions at astonishing, incredible speeds. However, looking at this from another perspective, we're facing a system that allows incredible profits without doing much of anything substantial.

For example, let's say we are two banks, and you are a bit slower. You work in milliseconds, and I work in microseconds. This means that one millisecond for you is a thousand milliseconds for me: I can do a thousand things in the time it takes you to do one thing. Let's say that in this fraction of time, I buy a thousand shares of company X. You, a computer, will register an abnormal increase in X's purchases and start buying that stock in the expectation of a price increase. But unfortunately I, playing in advance in the next millisecond, will sell you the shares I acquired a moment ago at a lower price, thus making a profit solely due to my technological performance. It's like stepping into a ring and fighting against a boxer a thousand times faster than you. That's why I wrote that today banks are willing to pay a lot to have their computers as close as possible to the terminals where financial asset prices are calculated. It's complete madness, unrelated to the daily lives and economies of ordinary people, a purely speculative game that allows banks to grab the money created by the central bank.

CHAPTER 6

The Failure of International Institutions

S.A. Professor Giraud, by evoking new global economic parameters, you implicitly call to mind to the role of multilateral international institutions. However, as Petrini emphasized when highlighting the gray areas of the food industry, they have neither great credibility nor great effectiveness. The war in Ukraine has taught us this . . .

G.G. On energy, sustainability and peace, the failure of the rule of international law, particularly the United Nations, is evident to all. And where the rule of law is lacking, a highway is opened up for authoritarian and violent tendencies. It's no coincidence that in these years, the myth of the cowboy has returned to prominence in the United States. This myth was constructed by the Republican lobby in the 1960s with a specific purpose: to prevent Congress from passing a federal anti-racist law. In reality, historically, the era of cowboys lasted barely two decades, from 1860 to 1880. Another detail that dismantles the macho rhetoric: A third of them were Black. Yet, in the films of John Wayne, the cowboy champion, Black people are practically nonexistent. We find the white man, his family, his home, his work and his gun. The message was: We don't need the state or the law, my gun is enough. And we certainly don't need supranational institutions capable of truly enforcing legal rules for everyone.

C.P. So far, in practice (which is different from declarations, which count for little if they remain on paper), international organizations have imposed an anti-community development model. Instead of fueling the mad race for increased production, these institutions should verify—and do so precisely and autonomously—that communities are genuinely operating in the interest of their food sovereignty and, therefore, socio-environmental sustainability and the common good. This would restore considerable prestige to them. It wouldn't even be so complicated to fulfill this mission, considering the budgets they have at their disposal and the collaboration that would be offered to them by local actors, who have every interest in ensuring that the places where they operate are preserved. Unfortunately, it seems that international institutions have taken a completely different direction, that of being complementary and therefore complicit in "business as usual."

The money to save the banks is always found in record time and without remonstrance. Silicon Valley Bank and Credit Suisse are glaring examples in this historical period. But even worse is the thought of the substantial sums that individual states invest each year in military spending. In 2021, in the midst of a pandemic, we reached a historic high: $2.1 trillion dollars.[1] In 2022, with the outbreak of the war in Ukraine, things certainly did not improve. Spending on the arms race by the United States alone amounted to $722 billion.[2] And other NATO member countries followed the same path. We are feeding war and destruction instead of feeding the world. To feed the world we would need €280 billion, which is about one-eighth of what was spent in 2021 alone on weapons.[3] Can we really consider today's world a healthy place to live?

2. A NEW WORLD

CHAPTER 7

First Civil Society, Then Politics

S.A. As we start a reflection on the role of citizens in the process towards a new social model, it is necessary to consider two key words: transition and sustainability. Are they synonymous?

C.P. I wouldn't want the term "ecological transition" to meet the same fate as "sustainability," a word that is now used by everyone and, I would say, overused.

We often associate the term "sustainability" with the verb "to sustain" with the aim of revealing how the goal of any company is to be supported and secure from an economic and operational perspective. But that's not the point. Rather, as I have already mentioned, the French are right to translate sustainability as *durable*. Any of our attitudes, political actions or ways of life must produce results, and these must last over time. It is a new paradigm that demands a change in our consciousness. If we accept this logic and apply it to our daily lives, we will break free from the mechanism of compulsive consumerism.

I believe it is necessary for "ecological transition" to be considered not a mannered or, worse, trendy concept, but rather for its true meaning to be understood: the opening of a new historical phase that could last for several centuries.

Since the birth of the Industrial Revolution, which emerged in old Europe before spreading worldwide, we have seen incredible changes in people's behavior, lifestyle, health and level of well-being. Extraordinary scientific achievements have been made. That process, initiated three centuries ago, has now reached a more than mature phase. As mentioned, one of the fundamental axioms on which the Revolution was based was that nature's resources were limitless. We have finally become aware of their finiteness. The Club of Rome intellectuals already warned us in the 1970s. Now that the finiteness is tangible, behaviors that continue to be disrespectful or indifferent are generating tremendous suffering worldwide. And, as Pope Francis says, the greatest toll will always be paid by the poor.

The ecological transition is the start of a historical process that requires us to be ready to question our way of life and the way we eat, produce, travel and form relationships. It constitutes a radical change that can only occur through a movement and a form of consciousness that challenges bad habits and fuels dialogue between different perspectives. Simply applauding "the ecological transition" and renaming a ministry is not enough. If a profound discussion does not truly take root in society, any institutional label risks being counterproductive or, at the very least, ineffective. We must, at all costs, support movements and associations to promote a campaign of awareness and responsibility with clear political implications. What I mean is that individual citizens must be made aware of the power they hold in their hands. Through their behaviors, they can influence markets and ensure that institutions also make more informed decisions. This will be the true politics of the future.

From this perspective, we could discuss how to reduce invasive monocultures or how to ban the pesticides that are poisoning us (the Veneto region is full of them, for example). If people can feel the weight of the issues at stake and realize that they can be active participants, the ecological transition process will be inevitable. Is it a dream? I think not, because fundamentally this is the authentic meaning of politics. Whether we like it or not, we will get there.

S.A. Will political or civil mobilization come first?

C.P. Unequivocally I'll say that politics comes after. The initial impulse can only come from civil society. Today, political debate on these issues is either absent or complicit. For years, we have witnessed joint sessions where the world's leaders acknowledge disastrous reports certifying the profound sickness of the environment and society in the face of a development model that, as it stands, will lead us straight into a wall. They listen to the alarms, study the documents, declare their concerns and then, in concrete terms, do little, very little, much less than what needs to be implemented immediately.

Ever since the 2015 Paris Agreement on climate, it has been this way. COP26 in Glasgow and COP27 in Sharm el-Sheikh were the latest striking demonstrations. An example? The delegations at COP27 once again committed to keeping the increase in the global average temperature below 1.5°C compared to the pre-industrial period. However, with the current measures adopted by various countries, it is estimated that the increase will be, already in this century, between 2.1 and 2.9°C, still relative to pre-industrial levels. To stay within the specified limits, it would be necessary to halve greenhouse gas emissions within the decade, an increasingly unlikely if not impossible goal to achieve. Words plot one path, actions another.

I also look at Italy: Politics lacks a medium-to-long-term vision; it thinks only in the short term. The electoral campaign for the 2022 political elections, in the midst of an energy and social crisis, was a concrete demonstration of this. I still don't see anyone fighting for a vision of the future. The heart of the debate perpetually remains that of tactical alliances and vetoes. But I urge the parties to be careful because they risk the disinterest of masses of young people reaching unprecedented levels. The data from the regional elections in February 2023 are proof of this.[1]

Let's return for a moment to the war and the pandemic. Two events that have brought back to the top of the global agenda the return to coal and nuclear power, as if the reflections on energy and environmental issues were worthless. Without considering, for example, that in the summer of 2022, many French nuclear power plants were shut down due to a lack of water.

Yet in times of emergency we continue to look to the past. The same goes for the wheat crisis. The conflict in Ukraine has a very significant impact because it involves not just any two countries, but two with a strong agricultural industry, known for the production and export of food, especially staples like wheat. In its initial phase, the war in Ukraine blocked exports and paradoxically (though not surprisingly, given the model of global agricultural production), the ones to suffer the first and most were countries in the global South, geographically distant from the conflict areas. About 50 developing countries, many in Africa and the Middle East, were importing 50% of their wheat needs from Russia and Ukraine.[2] This means that regions that were once the cradle of agriculture are now extremely dependent on Ukrainian wheat. A contradiction that we will no longer be able to support. We must overcome the paternalistic idea, which has proven profoundly wrong, that the West feeds the global South. In reality, it has only made these countries more vulnerable to crises.

I am convinced that we are entering a new historical phase. I believe there is a need for massive community and youth mobilization, of which I see the beginnings, to make it clear to political parties and those who control institutions that today the real politics is carried forward by civil society. A society capable of changing behaviors and bringing about a change that the political world will be forced to ratify. But it will do so at a later stage. Expecting that world to take the first step, to have a modern and rational vision, is not conceivable. The thinking of those who govern us focuses on the period from one election to the next. The movement for ecological transition will free us from a social model that makes us poorer and more defenseless, bringing about a change that will lead us to a happier life, as well as preventing us from sinking into the abyss.

G.G. I agree on the need for the political mobilization of civil society. Following Carlo Petrini, I'll also start from the energy issue and its connection to the epochal change we are experiencing.

To simplify, the 19th century was the century of coal, a raw material that was complex and costly to transport. This fact gave great negotiating power to the poorest social classes, especially those involved in extraction, processing and transportation. The social reforms of the 19th century, so innovative compared to the Middle Ages, are therefore closely linked to the coal economy. In the 20th century, the arrival of oil, much easier to transport, radically changes the scenario. The negotiating power of workers decreases significantly and Margaret Thatcher's "war" against miners in the UK in the 1980s becomes the tragic reflection of a new energy model. Oil, in essence, is the keystone of the success of neoliberalism, a system still dominant today but in clear decline. This is precisely because we have now understood that the resource of oil is not infinite and that its use has increasingly intolerable environmental and social contraindications.

S.A. Following the thread of your reasoning, a new era requires a new energy model, which will obviously have to be based on renewable and clean energies. Stating this seems almost too obvious. More challenging is understanding how it can become structural, beyond marketing and the enormous exposure that a word like sustainability has today.

G.G. There are two possible models. The first is linked to a political scheme that we could define as decentralized democracy. In this context, energy production is organized at the local level, say municipal and citizen. The formula that best fits in this case is that of energy cooperatives, based on non-profit, basic energy communities. A few years ago, attempts in this direction were experimented with in Denmark and especially in Germany. However, at that time, the conditions were not yet right, as the cooperatives were absorbed by for-profit private companies. But times have changed. And more than France or Germany, it is Italy, with its communal and municipal traditions, that boasts a social structure that allows for successful experimenting with this model.

Then there is the other option, contrary to decentralized democracy, based on the alliance between an interventionist state and a greedy private sector, hungry for profits, which would control energy resources from above, perhaps passing off dirty resources as clean. This is the Chinese model, but let me say, it is also the model that Emmanuel Macron's France has in mind. If we go down this path, a return to nuclear power, and more, will be almost inevitable. Another consequence will be that the cost of the ecological transition will be borne largely by the poorest, those who do not have the means to protect themselves from pollution and do not enjoy the economic profits deriving from the old, dirty technologies.

This is the time of choices. It is now clear that on one hand the conservative elite is fighting to maintain energy control even in the new paradigm of sustainability, while on the other, a progressive civil society is advocating for energy decentralization and democracy.

S.A. Professor Giraud, you mentioned the central role of cities. Do you really think that municipalities can be the driving force behind sustainability at the international level?

G.G. In Italy, you have the example of Bologna, which is at the forefront of these issues, as well as shared administration practices. Then there's the C40 network, Cities Climate Leadership Group, an alliance of 97 cities worldwide that represents one-twelfth of the world's population and a quarter of the global economy. It includes Milan, Venice, Rome and other major cities such as New York, Beijing, Berlin and others. Behind this movement, there seems to be a willingness to open up the possibility for cities to use public debt to achieve environmental and sustainability targets. If this were the case, the push toward integral ecology would be very strong, and neither national states nor international bodies could ignore it.

The political agendas of cities are decidedly more agile than those of states. This is closely related to land-use planning: It is possible to design small, dense cities without cars, connected by trains and integrated into an agroecological space that serves as a carbon absorption reservoir. This would also put an end to the highway system, a significant source of pollution and therefore a danger.

S.A. You referred to Italy, but our country is a clear example of the political failure of green and ecological parties, which consistently record very low levels of support whenever there are elections.

G.G. We can put France on the same level as Italy. This happens because, so far, in both contexts, the ecological parties have not built any global model. Their ideology, although environmentalist, is marginal because they fail to apply it at the systemic level. So, they talk about "green," but then they are very fragile when it comes to developing a vision that involves a change in the economic and social system. In Germany, where the *Grünen* have a real and visible impact on society, the phenomenon is different, although ambivalent. There is indeed a strong right-wing reactionary ecological tradition, rooted in the 18th-century rebellion against the rationalism of the French Revolution. In the second half of the 18th century, many

German intellectuals advocated a return to the land and nature in opposition to French rationality. This conservative and anti-modernist mystique constitutes the roots of right-wing ecology, especially in Germany but not only.

As we have said, there are two political options for ecological transition: One is authoritarian, its foundations based on the alliance between a paternalistic state and an ultra-liberal and hyper-capitalist private sector. A private sector that knows what it's doing, especially in conditions of crisis and institutional confusion like the one in which we are currently immersed. It is therefore necessary to pay close attention to the signals. Civil society is crucial in avoiding this tendency because, by nature and vocation, it pushes toward the "left-wing" option of transition, that of democratic decentralization. But we still lack a communication chain between civil society and politics. When *scuole di partito* (training courses organized by political parties) existed, the communication chain from societal values to politics worked. Now the gear needs to be rebuilt. Indeed, the first thing an ecological party should do is to organize schools to cultivate in the younger generations an awareness and knowledge of these issues.

C.P. I believe that precisely because a true popular ecological party does not exist, we must start again from the activism of the movements. In the United States, when conservative Supreme Court judges limited the government's ability to reduce emissions from coal-fired power plants in July 2022, a significant climate disobedience movement emerged. I want to quote from a document drafted on that occasion by American activists and also reported in some of the Italian media: "We desperately need a government that works against the destruction of the Earth instead of accelerating it, but petitions and pleas will not bring it about. Due to the power of the super-rich, the only option within our reach is to fight and bring society into 'emergency' mode by ending the fossil fuel industry; we must unite and do everything to shake people and raise awareness of the serious danger we are in: We must engage in climate disobedience. The tide can still turn, power can still change suddenly. All this, however, can only happen if a sufficient number of people join the fight."[3] This is where we must start. From the movements. As I mentioned earlier, I no longer have any confidence that politics can lead the way.

However, we in Europe have a civil, association-based, ecologist, environmentalist and socially sensitive world that has already set out on a journey. In this sense, we don't need to invent anything. However, we must tell this world that we can no longer wait. That the challenge is political and must be played out here and now. Clear and effective actions are needed from the grassroots. We propose two: changing the food system and establishing renewable and community energy systems.

S.A. In this regard, I believe it is useful to revisit the work of English activist and writer Rob Hopkins, author of *The Transition Handbook*. In *L'ecologia di ogni giorno*[4]

("Everyday Ecology"), Hopkins reflects: "In my meetings with groups, I notice that a significant number of participants do not come from activist backgrounds, they don't fit the classic profile. Many engaged in the transition would never advocate for this or that ecological cause. But is transition really a form of activism? Many participants think not; they believe they are simply reinventing and transforming the place where they live, nothing more or less than that. They act with the intention of improving the situation, not through a global campaign but locally, with action taken city by city, neighborhood by neighborhood, street by street. . . . It's an idea that can be seen as radical or simply as a way of doing what needs to be done. That's why many transitioners do not consider themselves activists. Moreover, the trademark of Transition lies in not settling for a single discourse—with which you either agree or you don't—but in addressing disparate issues like energy, finance, agriculture or transport. The multiplicity of subjects we deal with also plays in favor of such diversity: Someone who loves planting trees can join a transition project, as can someone interested in food growing, recycling or renewable energy. As soon as people like these identify with the movement, they easily move from one project to another and get closer to aspects they knew less about."

G.G. It seems to me a very interesting and correct point of view. The role of associations is certainly crucial, but alone it is not enough. Power dynamics matter; on our own, we count for nothing. We cannot change politics by ourselves. Other actors must be involved. We must aim for a broad alliance that breaks down the barriers of traditional activism.

CHAPTER 8

Protectionism and Food Sovereignty

S.A. I'd like to introduce another theme: Both of you often refer to "protectionism." Protectionism calls to mind statism and sometimes extreme forms of nationalism. However, both of you are openly affiliated with the progressive world: For this reason, this aspect of your thinking intrigues me greatly . . .

G.G. I don't consider myself a statist in any way. In my view, there are four social frameworks for organizing resources. Private property: A cup of coffee is a private good, and I am the owner, and no one else can drink it without my consent. The second alternative is the theory of public goods: Coffee is a public good, and the state tells me whether I can drink it or not. The third option is that of tribal goods, where the tribe or clan holds power. This is the case with Indian castes, but also with Italian mafias. We can locate the development of that framework in the reaction of the southern Italian nobility to the establishment of the unitary state: on the one hand, the statist perspective of the northern bourgeoisie, on the other, the tribal perspective of the privileged classes in the south. Finally, there is the path of common goods. Here a good example is the movement for water as a common good in Naples and the civic struggle that led to the re-municipalization of water in the Campanian capital. We mentioned it before. Out of that movement, legal scholar Stefano Rodotà developed his theory of common goods in 2012: There are goods that are neither private property nor state property but which reflect the inalienable rights of citizens. These are common goods, from the right to life to the basic necessity of water to online knowledge. Everyone can enjoy them, and no one can exclude others from the possibility of enjoying them. The point of impact of fundamental rights—and therefore the natural recipient of common goods—is no longer the subject but the person, a term that current jurisprudence is recovering as less abstract and more concrete.[1]

It seems to me that this is a very interesting perspective for Europe and for Italy. We need to focus public and political debate on the question: What are the assets, the resources that we should consider as common goods? For example, energy or health are certainly goods that should be managed from the perspective indicated by Rodotà. During the pandemic, the WHO spoke openly in these terms. Did anyone listen? No. Italy didn't, France didn't. No one did. Another sign of the failure of the United Nations and international institutions. Yet, the common goods approach would simply be a choice of great common sense and real social protection. The health of a family in Wuhan, China, is crucial even for me living in Italy,

France, the United States or anywhere else in the world. If we put a price on the vaccine, it means we are privatizing it. The result is that there will be poor people who cannot afford it and therefore the virus will spread through their bodies, nullifying the effects of the vaccine and compromising the health of the wealthiest as well. Free vaccines, the organization of health as a global common good, remain in the interest of the rich as well.

I came across a very interesting project in Geneva, in Switzerland, called the Drugs for Neglected Diseases Initiative (DNDi), established in 2003. Its mission is to organize treatments for diseases neglected by the private pharmaceutical sector because they are mostly prevalent in poor countries, where profit margins are very low or nonexistent. DNDi is able to distribute a treatment for hepatitis C in Egypt for €300, while the same treatment in Europe costs up to €3,000. A miracle? Absolutely not. DNDi has achieved this result thanks to the influence of civil society and the ability to engage in dialogue with and involve sovereign states and the Big Pharma private sector. DNDi is not a state institution, not a for-profit company, not an offshoot of a privileged elite; it is an organization that deals with health from a perspective of the common good. And not only does it work, but it works very well. It is clear, therefore, that health constitutes a universal but fragile good because privatization destroys it. This means that health is a global common good that we all aspire to share and its care requires the cooperation of all. I believe the same can be said for climate and biodiversity. The question is the following: What institutions can we invent to honor this status of global common goods? The United Nations can no longer hide the fact that it has become nothing more than a façade behind which the interests of the United States are hidden. Once American power was weakened by Trump, the United Nations became fragile. EU institutions hardly excelled when it came to saving exhausted Europeans from months of lockdown. The WHO was ignored. This suggests the need to think about some other institution: We need creativity. We need rules that allow all stakeholders to come to the negotiation table and engage in this collective insight that national states and representatives from civil society and the private sector should seek. At a philosophical, political and spiritual level, this will be the most important challenge in the coming years.

All the more so since the model should be applied to healthcare and energy, but also to other fields, such as culture and the economy. The latter might seem like a field that is off-limits, but that's not true. The tools are there and they work very well. Think about local currencies, like Sardex, created in Sardinia in 2009, which today boasts over 5 million transactions and over 10,000 B2B accounts throughout Italy. An incredible and extraordinary success story, yet still relatively little known.

Local currencies provide an alternative to the euro. Of course, the euro today is not a public currency, but entirely private, its governance held by private banks. Local currency initiatives are therefore a way for people and civil society to reappropriate monetary power with the creation of a currency as a common good.

Health, energy, economy. Even work should be understood as a common good. Founding labor policies on the basis of competition between companies and workers no longer makes any sense. Yet the majority of economists are still stuck on this point. In my opinion, the enterprise remains a common project, based on a community of workers pursuing shared objectives. Therefore, we need labor laws as a common good. There are experts, such as the French legal scholar Alain Supiot, who pursue this perspective, an interesting approach in which civil society plays an extraordinary role but without the state disappearing as a result.[2] On the contrary, state power assumes a decisive protective function to prevent common goods from being tribalized or privatized, as in the case of Airbnb, which from a fantastic idea of sharing has turned into a business that has destroyed the community life of city centers, condemning them to hit-and-run tourism in a wholly privatized logic. The same goes for BlaBlaCar, now a completely privatized business. If in the West the danger is privatization, the risk in Africa and other parts of the world is the tribalization of common goods. In countries like Mali, Mauritania, Chad and Burkina Faso, where the state is severely weakened, there is no protection for either private property or common goods, which are consigned to the violence of the clan that protects "its" resources according to tribal law. The same happens in Lebanon. Here, too, state structures are very weak, and a completely religious-tribal logic governs, meaning the death of civil society and common goods. All of this reasoning is to say that it is not true that focusing on common goods means thinking that state power is diminishing. That's not the point. It means affirming that it is the duty of the state to protect common goods without turning them into state property.

S.A. Is this the protectionism that we need?

G.G. The reasoning on common goods is necessary to precisely frame a term, protectionism, which, as you were saying before, has a cultural and political connotation that is alien to me. I'll try to pick up the reasoning from another point of view with an example: the carbon tax, which simply means introducing the principle that those who pollute must pay for the damage caused to the community. I would like Europe to protect its borders and its health through this measure. We need to tell the United States and China that if they want to trade with us, they must do so in a sustainable and ecological manner. Is this protectionism? Yes, it is green and social protectionism. Is this nationalism? Absolutely not, it's pure common sense. And we must do it now while we still have time. In ten years, we will no longer be in a position to reverse course.

C.P. To talk about protectionism, we need to talk about food sovereignty. So far, major international agencies like FAO and WFP (World Food Program) have almost always intervened based on an assistance and emergency principle:

supporting populations in food crises with transfers of food from abroad. This is a costly way of intervening and, above all, totally inefficient in terms of preventing food crises. Also here, as Giraud argued, an alternative exists: focusing on local productions according to the principle of food sovereignty. This is quite different from autarky. Autarky implies consuming only products from one's own area. Food sovereignty, on the other hand, says: I use and defend local products because they have a history and a culture, after which I look beyond my borders to obtain what I don't have. For example, it makes sense for Italy to buy coffee from abroad. But why get tomatoes from China, process them, put a tricolor flag on the can and then sell them in Africa, as we do now? This is how it works: Tomatoes arrive from China in the Agro Pontino. We can them and export them to Kenya and Tanzania at a lower price than locally produced ones sold on the domestic market. The result: Africans think they are eating Italian tomatoes and we engage in dumping in Kenya and Tanzania, selling tomatoes grown outside European health standards. This mechanism needs to be broken. How? By introducing an ecological tax on "fake" tomatoes.

The recovery of food sovereignty is an essential condition for building a more environmentally respectful and fairer civilization. When, on the other hand, continuous production increases are pursued through chemical fertilizers and pesticides, enormous monocultures or factory farms are generated, resulting in local areas being violated and disrupted, pushing a large part of the population elsewhere and compromising soil fertility and local biodiversity for decades. Never before in history have we witnessed the devastation of lands and aquifers as in the last decades, a devastation born out of the privatization of resources that should be common. Being sovereign, on the contrary, means entrusting to food communities the ability to decide what and how to produce, how to distribute and how to co-produce, involving the end recipients in this process.

S.A. Geneticist Salvatore Ceccarelli and plant breeder Stefania Grando, in their *Seminare il future*[3] ("Sowing the Future") explain what is meant by biodiversity and agrodiversity: "It refers to three distinct levels: diversity among species (for example, wheat, corn, rice, barley, etc.), among varieties within species and among plants within varieties. Consider that there are approximately 250,000 to 300,000 described plant species, of which humans have used about 7,000 throughout history for various purposes (food, medicine, textiles, construction, manufacturing, etc.). Currently, only 30 crops supply 95% of the global food demand, with four (wheat, rice, corn and potato) playing a predominant role. Furthermore, the number of varieties of species that provide the majority of our calories is much reduced compared to the past."

But there's more: "Let's ask ourselves: Who are the masters of food? It's certainly not us, even though we seem to have an embarrassment of choice, especially when we enter a supermarket. Whatever we buy, if we read the labels carefully, it's almost always a product from one of these food multinationals: Nestlé, Coca-Cola, PepsiCo, Kraft, Unilever, Kellogg's, Mars, P&G, Johnson & Johnson and perhaps a couple of others. Often it's not obvious because, for example, Nestlé, the global leader in processed food products (almost all packaged foods), in the UK alone sells under the labels of 31 different brands. But that's nothing considering that globally it controls about 8,000 brands, ranging from products that end up in baby bottles to those for dogs and cats. Leaving a supermarket without having bought a Nestlé product is really quite complicated." Are food sovereignty and biodiversity preservation therefore two sides of the same coin?

C.P. The protection of diversity is a guarantee of development. I'm not saying this; it's the law of nature. Everything—plants, humans, animals, even businesses and organization—is subject to the same law of evolution. We're born, we grow, stabilize and then decline. If we continue to pare diversity down to the bone, focusing on a few high-yield varieties, we risk finding ourselves in a world devoid of alternatives when these varieties start to decline. Speaking of diversity, I would add: Identities, both those strictly related to food and cultural identities, are defined only because differences exist. A neoliberal consumer society is anti-sovereignty and anti-protectionist because it must necessarily be homogenizing. Flattening differences is functional to a market model that wants us all to be consumers of serial and transnational products: all standardized and easy targets for marketing that erases diversity and sovereignty. The very origin of the expression "food sovereignty" testifies to this. In 1986, in Geneva, during a plenary session of the World Trade Organization, the elite of international politics decided to include primary production within the General Agreement on Tariffs and Trade. From that moment, choices regarding food production and trade, the environment, access to land and the culture linked to life in the fields would be subject to the neoliberal rules of the international market. In response, we saw the rise of worldwide movements linked to the farming world, with the goal of defending the authentic value of food, food not merely as a commercial commodity but a human right to be guaranteed and protected.

Starting from the rationale of civil society, in 1996 some international organizations who were gathered in Rome introduced the term "food sovereignty" into the international institutional vocabulary, using it to refer to the self-determination of peoples in choosing their agricultural policies so that they are in harmony with the ecological, economic and social fabric and guarantee access to healthy, nutritious and culturally appropriate food. Over the years, millions of farmers on all continents have corroborated the concept of food sovereignty. Our own Terra Madre network is a living expression of it. Ecuador enshrines it within its constitution (Article 281) as a duty of the state and the United Nations identifies it as a necessary precondition for achieving the strategic goal of "Zero Hunger" in the 2030 Agenda.

G.G. I am talking about economic and social

protectionism. The example you mentioned about tomatoes is very relevant. I would like to add an observation. In China, working conditions are not equivalent to those in the West, and this introduces competitive mechanisms that dramatically push down the wages of European workers. We don't talk about it much, but it represents a crucial aspect of the economic crisis we are experiencing. What should the response be? Very simple: buy exclusively ecological products from production chains that fully respect the criteria for decent work set by the ILO (International Labour Organization), as mandated by Goal 8 of the UN's 2030 Millennium Goals.

During my time as chief economist of the French Development Agency, I participated in a conference in Guinea Conakry. This country is a striking example of agrifood biodiversity, yet they consume almost exclusively Thai rice, and industrial production is focused on mineral extraction for European industry. I argued that this situation seriously endangers the country's food security. Representatives from the International Monetary Fund and the World Bank responded that Guinea's growth, and therefore its well-being, primarily depended on GDP growth and, consequently, more intensive mineral production. However, international institutions should serve a different purpose. For example, they should be funding and actually implementing the project for a Green Wall, an 8,000-kilometer [4,970-mile] barrier intended to prevent the southward advance of the Sahara Desert. This could be a foundational element of a new Marshall Plan for Africa.

It would simultaneously address the food security of a significant portion of the continent and provide the best response to the refugee crisis. And in Africa, this initiative could be supported by the renewable energy of the sun.

S.A. If it's really that simple, why has the idea of the Green Wall been discussed for over half a century, but we're still at a standstill?

G.G. Because the World Bank and the International Monetary Fund continue to impose a strongly neoliberal ideology based on low-cost flexible labor and oriented towards supporting markets in the global North. A voluntaristic project of this kind is completely out of the question for them.

C.P. In addition to this, two other phenomena work against eco-sustainable projects like this. The first is the widespread corruption among the African leadership classes, whose survival depends on the system described by Giraud. The second is the Chinese influence in the form of commercial neocolonialism and the conquest of territories to produce food for export and biodiesel or in any case to pursue maximum profit.

Africa boasts 60% of the world's agricultural land, representing a huge opportunity that has not yet been fully realized. The government of China, with its growing middle class hungry for food, has understood this very well. Not surprisingly, in the last five years, Chinese imports from Africa have recorded an annual growth of 18.2%.[4] However, the problem is that the trade in agricultural products has its limits in terms of volumes. And all this points towards the necessary industrialization of the African agricultural system, with all the negative implications that we are well aware of.

S.A. So, for you too, Petrini, it's a mistake to consider protectionism a "right-wing" concept, even though it is largely associated with protecting the borders or the rights of workers of one nation rather than another?

C.P. Of course it's a mistake; it's quite the opposite. Immigrants are necessary to protect our agricultural and gastronomic vocations. The problem is that today we pretend not to see that a part of what we consider our identity exists precisely because we can count on the presence of foreigners. Anyone who works seriously in the agrifood chain in our country is fully aware of the debt of gratitude we owe to those who come to Italy to find employment. Naturally, the correct balance is generated in food chains that do not lose sight of the common good. Only when agriculture is aimed at the well-being of nature and people do we avoid conditions of exploitation and injustice. Where, instead, food production becomes a mere tool for profit, competing unfairly with healthy agriculture, the rights of workers are denied and rules are not respected.

This mechanism primarily affects the most vulnerable and least protected individuals, and secondarily, the consumers themselves, who are offered low-quality products without any social and civil value.

CHAPTER 9

The Youth Revolution

S.A. Your call is a strong appeal to the mobilization of civil society. For it to have success and a future, it is necessary to involve young people. Do the conditions exist for this?

C.P. "If youth but knew, if old age but could . . ." We could sum up the relationship between old and new generations with this saying, widespread among the elderly in our countries. However, I believe that to better understand the dramatic situation in which we are abandoning our youth, the axiom should be reversed: "If youth but could, if age but knew."

For decades, we have exploited the Earth, its water and every type of natural resource without any limit; we have self-destructively violated our humanity, depriving ourselves of that sense of limit inherent in the nature of any living being, a sense of limit that determines its survival. We, "the old," have effectively taken the power of action out of the hands of our children. But now, signs of their rebellion are starting to emerge. Young women and men today perceive the tragedy to which humanity has exposed itself much more than we do. Consequently, they are willing to change ways of thinking and behaving, choosing more consciously and with greater foresight.

It won't be easy: The technocratic model that has dominated the West for over fifty years is having an equally alarming impact on the newest generations, including from a social and relational perspective. The constant emphasis on technological rhythms and rules has effectively led to the elimination of the concept of limits from the cultural substrate of the youngest members of society. I say this observing the sense of disorientation of many of the young people among the so-called digital natives, for whom technology—indispensable for many aspects—has created a kind of virtual world. Our intergenerational legacy, therefore, could result not only in a catastrophic environmental and climate scenario but also in the inability to pass down the essential tools for building a better future.

But not everything is lost yet. We need an intergenerational alliance. The more mature generations must make themselves available to young people to convey the importance of relational goods. Sociability, conviviality and living real and shared experiences will be the tools that nurture the desire for a fairer and more sustainable future inherent in every young person.

S.A. To be able to count on an active and purpose-driven civil society, however, formative opportunities are also needed to transmit the values of the environment, biodiversity and the community that Petrini referred to earlier. Are our universities equipped to do this today?

G.G. My answer is that, in most cases, university professors are not sufficiently prepared. On these issues, we see a very strong demand for knowledge from the new generations that does not find an adequate response. Professors of my age, around 50, often struggle significantly. That said, we are starting to see some interesting experiments in university contexts. Eco-friendly campuses, for example. I'm very familiar with one where I teach, at the University of Stellenbosch in South Africa, near Cape Town. In the Sustainability Institute, everyone is vegetarian, electricity is produced through solar panels and water is recycled, as is all waste. The building, of course, is completely ecological. Every morning before entering the classroom, students work together for 30 minutes in the food garden which is cultivated based on permaculture principles. This creates a powerful pedagogical atmosphere. There are other examples, not many, but they do exist. I can think of Schumacher College in Devon, England. Then there is the Transition Campus in Forges, south of Paris. Those who attend spend their days together, with a communal lifestyle based on "happy sobriety." In other words, vegetarian lunches, a kitchen that uses unsold food, simple cooking methods, the use of techniques like lacto-fermentation, recycled materials and permaculture practices. This communal way of life is the aspect that makes the biggest impression on the hundreds of people who visit the campus every year to attend courses or out of interest. We should multiply these experiences within our universities and invite politicians and bankers to experience for a few days the joyful and sustainable lifestyle practiced in those places.

CHAPTER 10

Revolution at the Table

S.A. Beyond the educational aspect, which mainly looks to the future, what can we do today, both young and less young, as citizens and consumers?

C.P. Each of us must enter into the mindset of aiming to eat 50% less animal protein, especially meat. No one can be forced to become a vegetarian, and I am not one myself. But halving meat consumption is something that can be done immediately. In the 1960s, when I was a teenager, Italians consumed an average of 40 kilograms [88 pounds] of meat per capita per year, and no one was dying of hunger. Today, we consume an average of 95 kilograms [209 pounds], and in the United States, it's even higher at 130 kilograms [287 pounds].[1] And yet, it is scientifically recognized that we reached the best balance between nutrition quality and caloric intake around the mid-20th century. Everything has gotten worse for us since then. When I say "us," I mean citizens of high-income countries because, for example, in sub-Saharan Africa, they consume only 5 kilograms [11 pounds] of meat per year. There, unlike here, quantities should increase in the short term, by at least three times. All of this is part of the protein transition, a significant aspect of the ecological transition we are discussing. Less animal protein in our diet means less waste, less energy and water consumption and less pollution. For example, in Italy we paradoxically import meat from Argentina and Brazil. The most emblematic example of this practice is undoubtedly bresaola, a salted and dried beef typical of Valtellina, where it has a PGI label, and other mountainous areas in northern Italy. The high market demand for this product, which to be met needs 25,000 tons of fresh beef every year, is supported by importing 95% of the meat. This comes mostly from Argentina and Brazil, where vast expanses of pasture are grazed by zebu, a particularly suitable breed due to its lean meat. Now, I don't want to demonize bresaola, one of the glories of Italian gastronomy; instead, I want to draw attention once again to the management of limits and the necessary process of changing eating habits that we should all be contributing to triggering. I repeat: Consuming less meat doesn't mean eliminating it from our diet; it means eating less and of better quality and buying it where there is a historical culture of livestock farming based on the characteristics of the local area. What needs to be banned is factory farming. In North America, I've seen places where 20,000 cattle are slaughtered daily—a chilling sight! The same happens in Argentina and Uruguay, markets where we source meat. These mechanisms are harmful and we need to abolish them. At the same time, in my view, special attention should be given to the issue of synthetic, cultured or lab-grown meat (currently there are different names for this product). This is the result of sophisticated bio-engineering techniques, multiplying animal stem cells in the lab to obtain meat similar to that of real animals, both from land and sea. Experts say that by 2030, this product will cost consumers the same as meat as we understand it today.[2] However, entrusting the protein transition to laboratory products would mean further handing over control of our stomachs to just a few actors. Behind lab-grown meat are the big multinational food companies (such as Cargill, Tyson Foods and Nestlé, some of which are also the biggest players in industrialized livestock farming and leaders in the global meat market), investment funds driven not by the desire to stop environmental decline but by the pursuit of significant profits and prominent figures from Silicon Valley. Many of these are the same who, in the 1970s, promoted the Green Revolution. Technology's assistance for our future is essential, but innovation also means knowing how to draw from the virtuous aspects of tradition (knowledge, practices), which become bastions supporting the defense of the human-nature balance. In many small-scale farms, animals are seen as a resource, not a source of negative externalities, as is the case with industrialized livestock farming. I'm referring specifically to situations where species don't spend their lives in confined spaces, fed GM-soy-based feed, but can live outdoors and meet their natural behavioral needs. Here, livestock farming often develops alongside crop-growing, forming a closed-cycle system of matter and energy, where no waste is produced, the balance of ecosystems is preserved and food safety is ensured. This is the direction we need to take in the future.

G.G. Great care must also be taken over fish. The seas and oceans are being depopulated and, in the future, could become water deserts. It's a terrible effect of climate change. And do you know what neoliberal economists tell us? That it's not a problem because fish can be farmed, and this would have a positive impact on GDP. Even more absurd. Can you imagine a sea without fish? It's a question of civilization. We need to support a broad international debate, at the intergovernmental and United Nations level, to ensure that the seas and fish become a common good and are protected.

C.P. And here we return to the theme of education. Because we are trapped in a food system that, on the one hand, turns the seas into deserts, especially due to trawling, particularly practiced in Latin American countries to produce animal feed, but on the other hand, leads us here in Europe to always consume the same fish: sea bream, sea bass, salmon and swordfish, for example. The consequence is that less-valuable catches are discarded. The same happens with meat. We can't always expect to eat filet or sirloin, and fortunately, today even top chefs are starting to propose more offal.

Think about this. Today, the famous chefs who have become television stars and guide consumption are practically all men. But the history of cooking is actually a female story, a story driven by poverty that has developed in a community-based, peasant context, where waste was neither possible nor acceptable. A typical dish from Tuscany is ribollita, a soup made with stale bread. Then there's lampredotto, prepared with cow's fourth stomach. In Piedmont, we have ravioli that were traditionally filled with the week's leftover meat. Selling, increasing, accumulating: In the name of the capitalist economy, we have set aside our traditions and our culture. But agriculture doesn't work according to the logic of capitalism. We need to recover the subsistence economy. This would allow young generations to return to enjoying a food culture, and all of us to eat better and enjoy better health. If you think about it, no child knows the origin of the food they consume. Try asking them where the fruits and vegetables at home come from. They might think that radishes grow on supermarket shelves.

S.A. How do we overcome the culture of waste and discard on which Pope Francis has also extensively reflected?

C.P. First of all, it must be said that we are not starting from scratch because in the last 15 years, giant steps have been taken both in terms of skills and awareness. Is it enough? No, it's not. The numbers are clear, and it's worth repeating: Even today, about 30% of all the food produced globally does not reach anyone's table.[3] Just under a third of this waste can be attributed to domestic behaviors. The rest, however, is lost in earlier stages: in the fields or during processing, packaging and distribution. Clearly, waste is not an accident but an indispensable element for a certain economy. Without throwing away a significant portion of the food it produces, the system cannot continue. It cannot continue without wasting, without polluting, without making people sick, without causing damage. We can get this "dirty" food only under these conditions.

So, what to do? First of all, we need to become aware and feel responsible in the face of reality. We must say it loud and clear: Any citizen, all of us, can choose to live without waste. It's up to us to be part of a small revolution: the first step towards big change. A change that is based on the rediscovery of a principle that's fundamental for someone like me, a gastronome: the principle of pleasure. After all, it's almost banal to say, we are what we eat. Carefully studying how food is produced and marketed is something that would open anyone's eyes. It allows us to understand the complex mechanisms that govern the production world and, consequently, our own lives, but it also allows us to imagine and take different paths to build the foundations of a sustainable future. I think, act and write as a gastronome: My assertion, therefore, is of a fundamental right of humanity, the right to pleasure, which remains a natural, physiological right, the denial of which has been one of the elements that has led us to the situation in which we now find ourselves.

CHAPTER 11

Community Economies

S.A. In the first chapter, Giraud highlighted the perverse mechanisms that drive banks to defend the neoliberal paradigm. On this front, what is the space for action for individual citizens and civil society?

G.G. There is space, and it would be revolutionary. Virtuous banking models like ethical and cooperative banks, which make sustainability a real asset, do exist. Imagine what would happen if all savers shifted all their assets and investments from traditional banks to ethical banks. It would change the world. Starting from the ground up, we would change the face of capitalism.

C.P. This is a true political perspective. Local, ethical and cooperative banks should be given a political mandate to support the ecological transition and community economies, starting, for example, from groups who want to form energy communities. Citizens would thus have one more reason to trust these banks.

S.A. I've frequently encountered a word that marks the perimeter of the integral transition, and it has been explicitly or implicitly recurring in this conversation. The word is community. What does it mean for you?

G.G. Community is essential because the neoliberal culture and economy that are leading us to disaster thrive on individualism. Without community, we won't be able to escape the trap. The world of yesterday was caught in a vise. On one side was the communist ideology that dreamed of a total revolution, so vast that it was practically unattainable because no one ever imagined and promoted the steps that should have led to that revolutionary change. There was no first step, second step, third step and so on, only the chimera of the final result. And an endless wait that marginalized to insignificance any attempts at reform and civic engagement. This model was opposed by neoliberalism, an economic ideology based on debt, in whose vision the future coincided with the balance owed by many to the elite of a few creditors. A way of saying that, essentially, the future does not exist. Also on this front, there is no room for community initiatives.

C.P. For my part, the expression "food community" was created in 2003. I needed to find a binding term that would define and identify those invited to the first edition of Terra Madre. That wording allowed us to effectively focus on our target: people and organizations who shared some common values, but above all, who put food at the center of their lives, which is absolutely not something to be taken for granted. The meaning of "food community" has gradually transformed into a holistic and systemic concept. From the initial idea of only food producers, the concept has expanded to also involve consumers who choose to support farmers. In this way, consumers become co-producers. They are individuals aware that eating is an agricultural act, a community act and, ultimately, a political act. They are part of the community because they show awareness and responsibility. They refuse to be passive subjects who do not pay attention to what they buy and eat. They consider food as the goal of a cyclical production system because they are part of a larger system, the natural one. Being co-producers means being part of a community, along with those who cultivate, farm, process and distribute. It means sharing virtuous actions and ideas for the rebirth of a world that can again be harmonious and balanced, honoring the Earth's prosperity and regeneration.

Ultimately, food communities are the expression of an identity that remembers the past and has a clear vision of what it wants for the future. But watch out: Food communities and, more broadly, sustainability and ecological transition communities, cannot be mere rhetorical exercises. They must immerse themselves in their local areas and transform into economic models. There are plenty of examples. I'll choose one that has developed with significant numbers in the homeland of neoliberalism, the United States, an extremely interesting community laboratory: the organic farmers of Concord Commons. In 1998, some residents of the Chicago area and biodynamic farmer John

Peterson founded the Angelic Organics Learning Center, which, through Angelic Organics Farm, aimed to help people reconnect with the land, farmers and their community.[1] The farm is now one of the largest in the country and welcomes thousands of families from Illinois and Wisconsin who gather there to buy agricultural produce. In 2005, they also started an educational program in collaboration with Concord College, a community garden in Rockford, northwest of Chicago, which includes the Concord Commons Youth Garden, supported by volunteer families from the neighborhood. The program provides technical and scientific assistance for organic gardening and youth education. Once a week, children meet with the volunteer coordinator and staff to decide on activities: sowing, fertilizing, watering or harvesting products that are then distributed to neighborhood residents, used for community dinners and also sold at the local farmers' market.

The explosive strength of food communities lies in their very nature. By becoming consumers of food, we are actually letting ourselves be consumed by it. As we have seen, food eats us because it eats the land, its resources, its regenerative possibilities. Zygmunt Bauman explains it well: "Whoever is part of the community of consumers is, in turn, a consumer product."[2] The same happens with food for those who decide to follow the system; it is sad to recognize this, but that's how it is. Food communities (and the same goes for energy communities) escape this fate because they are active social bodies: They eat food instead of being eaten by it because, being co-producers, they take care of nature. The extraordinary thing is that each of us can participate in a community. You just have to want to.

This community structure needs a common feeling. And once again, we are shown the way by Francis of Assisi, whose words, still alive after eight centuries, teach us how we must start by doing what's necessary and then move on to the possible before finally discovering that we can do what at first seemed impossible. The first step is to ensure rights that are not currently guaranteed in much of the world: the right to food, the right to decent work, the right to live in a clean and healthy environment. The second phase is that of the possible: Eating 50% less animal protein is possible here and now, as is transferring our savings to an ethical bank. The third step, the impossible, will come on its own: a truly sustainable global policy, world-level institutions capable of defining and protecting global common goods and so on. Underpinning everything is the construction of a new community model.

G.G. Allow me a comment on Petrini's words. If you think about it, in the global and globalized perspective in which we are immersed, technically common goods that by definition cannot belong to individuals cannot even be state-owned because that would require a global state, a global administration to take care of them. The community way is the only one that is realistically and pragmatically viable.

C.P. It's a challenge that opens up a purely economic perspective: Food communities are the starting point for deindustrializing the global food system. Everything we have seen so far represents the failure of a top-down model. Developing an alternative from the bottom up instead responds to the need to combine ecological reasons and economic criteria. The two things do not exclude each other, and in fact are complementary. Let's start with words. "Economy" and "ecology" have a common root. In Greek, *oikos* means home, *nomos* means standard of conduct and *logos* means rational thought. Combining these terms, it is evident that if economic criteria are not guided by ecological and rational thought, we cannot achieve good housekeeping of our common home, nature, Mother Earth. The ecological perspective leads us directly to the valuing of local economies, which by their very nature make sustainability one of their cornerstones. Just think about distribution networks. Food produced and distributed within a local radius will certainly be less impaired by chemicals, fertilizers, pesticides and preservatives because it does not need them, as it is being sold within a community. The same reasoning applies to local livestock farmers, who will avoid polluting the land with antibiotic-contaminated manure, ammonia and other substances typical of large factory farms. In other words, local economies survive only if they preserve the biodiversity of the local area and its crops. The seasonality of products is not only a matter of taste, but becomes an indispensable prerequisite for productive efficiency. All this benefits natural cycles.

In this framework, another decisive element comes into play: the participation of the community, which even just to safeguard its health must verify that there are no negative repercussions on its local area. This significantly reduces the possibility of polluting and unsustainable production processes being initiated. So this is what's needed: control and responsibility. Not an induced responsibility, but one that is sincerely felt and lived, as the population naturally takes charge of managing its local area, becoming an active part of its development beyond the simple act of buying and consuming food.

One last element: We are not thinking of isolated communities separated from the context. Local economies need to interact with other economies, otherwise they will fail. The model remains that of a network, not competition. This ensures the exchange and knowledge of effective examples in specific contexts that can serve as partial or whole inspiration, even in other areas or for other productions. It is on these foundations that we have built the global Terra Madre network. Today, Terra Madre can be seen as a 'multinational' of local food. And in speaking of a multinational, I am not tripping over a contradiction because, in this case, it is being steered by millions of people scattered in every corner of the world, growing crops, rearing animals, caring, practicing and eating in the name of the common interest and not for the profit of a few.

S.A. Let's go to the root of this reasoning: Does the ecological transition imply a different conception of the principle of private property?

G.G. Private property is a temporary concession given to humans to facilitate their time on Earth. The values we experience today—freedom, equality, fraternity—are part of a program of political philosophy. But if fraternity remains tied to private property, it doesn't work. A simple example is offered to us, once again, by the coronavirus pandemic. If we privatize healthcare and put a price on the vaccine or tests that detect the infection, the poorest cannot afford them and the virus will continue to spread. Genetic mutations will occur and vaccines will no longer be effective against the variants, which will return stronger than before. This means that even the richest have an interest in ensuring that the poor have access to the means to counter the virus and that health is not treated as a private good. These are the clear limits of private property. Similarly, it is evident that nowadays private property is part of the ecological problem. In the United States, if a person finds oil in their backyard, no one can prevent them from exploiting the resource as they please. But if we believe that the climate is a common good that belongs to everyone, then we need to limit the right of private landowners in the United States or any other country to exploit the land. This would mean acting with discernment. However, as we have seen, in Europe, the United States, Japan and perhaps much of the world, for about 40 years we have been living in a form of neoliberalism that attempts to absolutize private property. Today, some ideologues, including economists, aspire to the global privatization of goods such as the genetic code, the indigenous knowledge of native Amazonians or women's bodies, for commercial purposes. Everything can be transformed into a patent and into merchandise for sale. This is the phantom of neoliberalism, and the pandemic has clearly shown that it leads to catastrophe.

Fortunately, new generations, or at least a part of them, are already in some ways attracted to a more attentive and responsible consumption model. I repeat: This doesn't apply to everyone and there is still much to be done, but this is the case for many young people. Before buying a plane ticket or a bottle of water, they think not once but ten times. This is a useful and necessary attitude. A European citizen, if they decide to become vegetarian, to recycle waste, to stop flying, to stop driving vehicles with internal combustion engines and to live in a thermally sustainable building, can reduce their carbon footprint by at most 30%.[3] Even if it's not enough, it's still a lot. The remaining 70% is determined by factors that do not depend directly on individual behavior: large-scale infrastructure, businesses, state policies. However, impacting even that 30% is important. And it has nothing to do with a return to the past. On the contrary, it represents a response to the anxiety of those who think that we are powerless in the face of disaster. No, we are not. We can do a lot at the individual level and even more at the collective level. European countries have set a very clear goal: net carbon neutrality by

2050. Will we get there? We must get there, we must work to achieve the goal.

C.P. When I refer to a new subsistence economy, I'm not thinking of a non-monetary exchange model based solely on the consumption of locally available natural resources or little more than barter. Nor do I imagine a closure to external contributions that leads to autarky for states and, for individuals, to the grim life of past centuries, especially in rural areas. Instead, I argue that the objective can no longer be accumulation. The goal must be to guarantee the right to education for children and a period of vacation for the family. What some fail to understand is that the subsistence economy pushes us to sharpen our wits and optimize energy and resources, thus avoiding waste. This way, the true purpose of food production is restored: food being produced, first and foremost, for nourishment, not to be sold or, worse, hoarded. This vision may seem limiting compared to the consumption dynamics we are used to, but it actually allows the local economy to be strengthened and generates positive repercussions for the entire community. This stimulates the active involvement of all citizens, promoting sociality and ensuring the defense of the soil, the landscape and our historical memory, as well as encouraging diversified diets that follow the seasons and provide nourishment in a healthy, delicious and culturally appropriate way. Finally, local production also defends biodiversity, the vital resource of every community. In summary, revaluing the subsistence economy means valuing good living, not wealth for its own sake.

In one of our dialogues, Pope Francis recalled a phrase dear to his grandmother: "The shroud has no pockets." I responded by telling him what a formidable elderly cook from a village in my area once told me. She ran a highly regarded osteria but only opened for lunch. Many couldn't understand why, when she could have earned much more by opening for dinner too. "I don't want to be the richest in the cemetery," she would reply. So, in the evening, instead of working, she spent time with her family while preparing preserves and sauces for the next day's service, giving continuity to important traditional knowledge.

G.G. We could define it as an economy of flow, as opposed to the capitalist one, which is an economy of stock. The stock economy is based on calculating how much what I have today can yield me tomorrow. The value comes from the sum of the profits I imagine I can make. Flow, on the other hand, looks at the value that something has for me at a particular moment. The fundamental principle is the energy necessary to live (well) here and now. Unfortunately, today all finance is based on the false idea of the need to capitalize.

C.P. Local and community economies are inspired by the logic of the flow. Here is where our reasoning on the economy and the leadership of civil society converge. Because it is in the local economy that movements and active people have the possibility to truly make a difference. In the globalized economy, we are nothing. Or, even worse, we are numbers subjugated to compulsive consumerism in which we are constantly seeking an unattainable happiness. In the participatory economy and in food sovereignty, on the other hand, the community plays a central role as a lever for development, including economic. I want to introduce a concept that is very dear to me, that of creative idleness, developed by my friend, the sociologist Domenico De Masi. If we move away from the mystification of accumulation at all costs, then we begin to see the possibility of building mechanisms that, alongside the production process, respond to people's educational and cultural needs. If we reduce the time dedicated to work, there will be more time for reading, art, creativity, theater; in other words, more time for building a horizon of meaning for our lives. This is what generates happiness! Don't spend Sunday afternoon at the mall, perhaps even without being able to buy anything. Creative idleness allows me to reclaim time and make it functional to my happiness. It also becomes a driver for a new demand for work, thus generating employment in the fields of the visual and literary arts we were talking about, but also in that of local agricultural and food production. If I have more time, I can afford more frequent visits to farmers' markets or nearby farmers, not just to buy their products but to learn about what I'm buying and the characteristics of that particular food.[4]

Based on these reflections and this new demand for healthy and sustainable products, in the early 2000s farmers' markets began springing up in the United States—yes, in the USA—before developing in France and also in Italy.

The entire fair-trade sector also continues to perform very well economically.

S.A. What impact can this kind of conception have on a strategic sector in Italy such as tourism?

C.P. An extraordinary, revolutionary impact. We are in the midst of a historical fault line, a contradiction that must be resolved. Because despite social impulses that have given rise to significant experiments in alternative production and distribution models, through phenomena like economic centralization, urbanization and deruralization we have depersonalized the small towns and historic villages of our beautiful country. We must also reflect on the implications of this, as we register a truly alarming cultural impoverishment and loss of memory.

First and foremost, we must distance ourselves from that type of dialectic, which by now has become rhetoric, that labels anyone attempting to preemptively consider the effects of unlimited development processes as nostalgic for a bygone era. I say this with full awareness of how harsh and often cruel life still was in the early 20th century in the countryside and rural areas across Italy. Yet I am also convinced that economic progress should not lead to the extinction of cultural values and effective social structures that have been deeply rooted in local areas for centuries.

The role that villages have played in history is recognizable through the artistic, architectural and scenic richness they still hold today, even when uninhabited. This is where the cultural biodiversity of our country was constructed, in the thousands of small societies founded on local economies, capable of expressing their uniqueness in a completely independent way. They are the many little pieces of a precious mosaic that has always distinguished and characterized Italy to the rest of the world. However, in recent decades, we have seen thousands of villages lose their populations and be left unrecognizable. In the best-case scenario, they are transformed into dormitory towns, in some cases tourist destinations that have lost their identity or even into ghost towns. There are several causes behind this process, but I believe that among the main culprits are large-scale retail, the arrival of e-commerce and the spread of mass tourism which cares little about the needs of the few remaining resilient local communities.

By allowing these expressions of the capitalist model, which tends to centralize human activities and channel profits to a single point (often identifiable in a single person) to establish themselves in the socio-economic fabric, we have gradually allowed the disappearance of workshops, artisans, osterias and grocery stores and the abandonment of small villages even by priests and religious communities. In short, almost unknowingly, we have rendered an immeasurable heritage sterile and soulless.

This certainly applies to places that have been abandoned. But it also applies to villages that attract a broader touristic interest and have been transformed into what are effectively open-air museums. Deprived of an indigenous community, residences have been absorbed into the holiday-home circuit or turned into hotels, osterias transformed into Michelin-starred restaurants and useful shops into aseptic souvenir stores.

Thus, even when we identify tourism as the panacea for all ills, we must remember the potential side effects it can generate. Healthy tourism, which does not want to produce negative externalities, knows how to manage its limits. We can no longer measure the efficiency of tourism solely in terms of the number and flow of people. This logic becomes counterproductive because it leads to privileging and increasing services and benefits for visitors without concern for rebuilding a genuine socio-economic fabric. If we do not understand this, we lose the essence of tourism as should rightly be understood.

Let me add this: A lively and happy community, ready to host and welcome, capable of expressing identity traits and cultural aspects, should and must be recognized as the ultimate tourist attraction. If instead we continue down the same path we are now on, we will see a definitive disappearance of life from Italian villages.

Tourism as it is still understood today can be purely economically advantageous but is not sustainable when measured against the authentic value of durability. We need to start by monitoring its quality. Until a few decades ago, those arriving from afar could immerse themselves in an experience made up of the village bakery's scents, the sound of bells, high-quality artisan products and the hustle and bustle of osterias, which served not only as ambassadors of gastronomic typicality but also as meeting places between locals and non-locals. Responsible tourism must return to emphasize this type of hospitality before memory fades, small centers deteriorate and identity traits are lost forever.

For our country, tourism, as well as agriculture and the dissemination of the culture of marginal places, is an essential opportunity to subvert systems that make the human presence on our planet hostile and fragile. Approaching these activities from a different perspective, abandoning the logic of accumulation and adopting responsible behaviors are the winning strategies to bring back and revitalize local economies to the surface, to rediscover the identity traits of our provinces, to return to a healthier lifestyle in harmony with natural cycles.

Today, we enjoy possibilities that would make life in villages much happier and more harmonious than in the past: new technologies, energy communities, ethical purchasing groups and all kinds of ways of communicating and creating connections. Providing the right tools to what I call resilient communities also means laying the foundations for restoring the relationship with nature and ecosystems. A new balance that would truly enrich the lives of all living beings in a local area. That's why I am convinced that restoring dignity to marginal places and small towns in our beautiful Italy is an innovative, sustainable and visionary practice par excellence.

G.G. The question then becomes: What tools do we use to strengthen local economies? Speaking of innovative financial instruments, we have already referred to Sardex, saying that the goal we must set ourselves is to develop social, local and complementary currency systems. Monetary sovereignty is essential to develop the local economy. The euro, being controlled by national private banks and the European Central Bank, cannot fulfill this function. However, local currencies have a significant sustainability problem in the long run. One possible solution could be that each municipality, or more generally each local authority, could decide to pay part of the salaries of its administration and collect taxes in the local currency. This would close the currency circuit and ensure the long-term existence of these instruments. Of course, banks would never accept such a proposal. The formation of an international network of local currencies linked to public administrations would create real competition with the euro and probably weaken it, creating significant disadvantages for large globalized credit institutions. For this reason, individual initiative and pressure from public opinion would be decisive.

S.A. What is the role of digitalization in this process of building community economies?

G.G. The transition we are imagining and building cannot ignore the digital revolution in which we are fully involved. I believe that the use of the web and social media as it is currently organized and regulated, or rather, currently unregulated, is one of the greatest obstacles for those who want to promote the community principle because the model it is based on is that of the tribe or—to use a more fashionable term—the bubble. And the bubble is a closed, self-referential and defensive circuit. It is very far from the perspective of the community of destiny for which we should be aiming. The massive way we use the internet has another deadly consequence: deterritorialization. If I spend a large part of my day online, I am not engaging with the place where I live, I am not part of real communities. I live in imaginary bubbles that do not exist. If we think about it, it is largely thanks to the internet that the Marxist concept of the proletarianization of labor survives. For Marx, the worker was "lost" within the power of machines. The same thing happens in the digital world. Except for a few rare cases like Wikipedia, where the public is also a contributor, the web user is a passive subject. And, in general, the society that the large digital multinationals want to build is a passive reality made up of small, uninfluential bubbles.

C.P. It is the capitalism of influencers, where consumers are the new proletarians and platforms with their influencers are the new capitalists.

G.G. That is why a digital transition is also indispensable, an internet 3.0 based on decentralized and cooperative models.

C.P. Community is participation. This is why I am interested in new forms of association like Fridays for Future (and fortunately, they are not the only ones); they use digital tools effectively but also physically engage, take to the streets and advance political demands for change and disruption. In this way, shared action practices are created, and one of the main characteristics of communities, emotional security or emotional intelligence, is generated. Unlike individual actions, where whoever makes a mistake is considered guilty, when a community sets out and makes a decision, it knows that it may succeed but it may also have a negative outcome. In the latter case, no one is singled out; instead the mistake is analyzed and considered something to be studied to improve in the future.

The second pillar of the community, alongside emotional security, is austere anarchy. It means that I cannot go to Chad or the Amazon and expect my membership-based associative model to adapt to those contexts. Specific characteristics must be respected; a community network must value diversity. The network of 6,000 Terra Madre communities works precisely because it applies the principle of austere anarchy.

G.G. An important point is the communication model determined within a community. The traditional conception where the leader, typically the father, speaks "the truth" and everyone must obey is typical of a tribalized context and unnecessary and harmful patriarchalism. What we need instead is synodal spirituality that fosters participatory dialogue within communities.

C.P. Charisma, however, should not be stifled; conditions must be created for those possessing it to serve the community, otherwise the communities themselves become centers of power.

G.G. This is not an easily achievable goal. We Jesuits talk about communal discernment, but it's one thing to talk about it and another to implement it in daily practice. There's another crucial point: For the community to function, it must be open. Here, the keyword becomes hospitality. Hospitality that must be modeled with respect for the uniqueness of each community, the austere anarchy Petrini talks about.

C.P. I believe that Slow Food's main mission is the protection of biodiversity. That's true wealth. This principle should also be applied to communities. Everyone in their diversity does what is most in line with their nature and local area, but they do it in an austere spirit, always aware that there is an overarching way of thinking that, to some extent, sets the general guidelines.

3. TOWARDS A NEW SPIRITUALITY

CHAPTER 12

The Happy Transition

S.A. From a communication perspective, the environmental and social transition is often associated with concepts like degrowth, sacrifices and unemployment crises. Is this really the case?

G.G. In 2022, I published a report on the energy future of France up to 2050. That study showed that a million jobs can be created solely in agroecology in my country. In essence, less oil and less coal mean more employment. In the 19th century, slavery was abolished precisely as a result of the energy revolution first of coal and later oil. If we now decide to give up fossil resources, the capitalist system will once again face the problem of how to find low-cost labor. We must, therefore, align the transition to green jobs with a strong strengthening of workers' rights. And it's a perfectly feasible choice. Thanks to technological progress, unlike two centuries ago, agriculture can be done with much less effort and without exploiting workers. Another decisive theme comes into play here: free access to new technologies, the so-called copyleft (making a work part of the public domain). Innovation must be made available to everyone. The work of those operating in this sector must certainly be rewarded, but the principle of international common good must be introduced through common license mechanisms.

S.A. Eliminating intellectual property rights means reversing wealth calculation mechanisms. And here we come back to the issue of GDP . . .

G.G. GDP is a stupid convention, as is becoming increasingly evident today. We must consider that in the 1930s it became a fundamental parameter to measure a nation's ability to wage war. Germany, in particular, was under scrutiny at that time. Those were the years when Adolf Hitler and Nazism came to power. This explains why, for example, women's domestic work does not generate GDP, because it is not instrumental to war. Today, however, we need to measure the sustainability of societies and economies and their ability to maintain peace. GDP is no longer useful to us. It no longer represents reality. Like the unemployment rate, which we have seen also does not represent reality. In fact, the neoliberal regime has created a society of unemployed people, and nobody knows. I'll go further. Most employment is determined by what the Americans, following the anthropologist David Graeber, call "bullshit jobs," jobs with no meaning. Whether you perform them or not, nothing changes for your community. They are useless, often bureaucratic jobs. The task of the ecological transition is also to restore dignity to work by generating real and useful employment.

C.P. The centrality and exclusivity of profit are now obsolete. We must think about and value common goods and relational goods. And I'm not talking about theory. Relational goods are tangible goods. Think about the distribution system. The regulatory favor enjoyed by a company like Amazon has corresponded to the systematic desertification of local shops, farmers' markets and so on. It's a distribution model that annihilates relational goods.

In the digital and delivery platform bubble, the beauty and architecture of places disappear. Like the smells and meetings that could happen when buying a wedge of cheese in a shop, an encounter with the cheesemonger but maybe also the priest or local elders. Transforming villages or neighborhoods into dormitories means losing the spontaneous sociality and cultural biodiversity that have always characterized our peninsula. Amazon's perspective, on the other hand, is that of dormitory villages, without visitors and without humanity. Villages that are good to put on Airbnb, with no culture and no relationships. Giraud spoke of good employment. Proposing in a modern and technological key the model of small local shops means creating good work and social networks through which the exchange and transfer of knowledge and vocations between generations can be kept alive: This is the difference between capitalism for the few and a participatory economy.

The new proletarians are born exactly in the context of the new digital capitalism. If the highest ambition we can propose to a young person is to become an influencer or a content creator, then we are responsible for the proletarianization of our future. The use of the internet—meaning how we use it, not the tool itself—is partly responsible for the impoverishment of our communities. We must become aware of this because it is an important element of the paradigm shift that we must generate.

G.G. Social media was born in 2004 with Facebook. That's less than 20 years ago. It's a young market in need of regulation. We cannot think of leaving it in the hands of a bubble. The consequences, and we're seeing them, would be very serious. Think first about what Donald Trump did and what Elon Musk is doing now with Twitter. And what if I said that Twitter should be shut down? Why not? Because it can't be said? Of course, it's a provocation, but the question must be raised. And it must be raised for Instagram and all the other social platforms. If these media become tools of propaganda for illiberal regimes or neo-Nazi movements, how should we react? I believe that in situations of serious social danger, limitations and even bans should be introduced. We now have the paradox that many top executives of the big five tech companies (Google, Apple, Facebook, Amazon and Microsoft) forbid their children from using the internet. Essentially, they're saying: "The internet manipulates minds and alienates people; this stuff is not for you." It seems to me a striking proof of the system's illness. But it's not necessarily the case that we have to continue like this.

C.P. However, we must be careful not to fall into the deceptive opposition between innovation and tradition. Today the dominant tendency is to keep them separate and see them as antithetical, with traditionalist conservatives on one side and progressive innovators on the other. This schema doesn't exist. Starting to dialectically integrate the two dimensions is a first step in building a different future. Because, paraphrasing Oscar Wilde, true innovation is a well-executed tradition, and if a tradition does not incorporate innovative elements, it is destined to die. Reading instead of spending time scrolling through social media, or shopping at small, locally owned shops, perhaps by placing an order online, is the most modern thing one can imagine. It is a matter of great active freedom, not passive. And the ecological transition is an act of liberty.

In 2014, I wrote a book with Luis Sepúlveda—a dear friend whose memory I cherish—titled *Un'idea di felicità* ("An Idea of Happiness"). The conclusion was that happiness is achieved when you are an active subject. If I don't matter, I can't be happy. And if I don't participate, I don't matter. So, social participation is the lever of happiness.[1]

S.A. In what sense is it a lever of happiness?

C.P. I'll answer using Sepúlveda's words: "Throughout the long journey of my experiences as a writer, as a journalist, as a playwright and, last but not least, as a citizen, a fundamental idea has always been present in me: Everything that is done for a better world has a starting point, and this starting point is to conquer the right to a full existence. An existence that is happy, in the most complicated sense of the word. Because the term 'happiness' implies many things. Knowing, for example, that someone close to us is living in a situation of social injustice is an injury to our idea of happiness. It is therefore in the name of that idea that we are working when we contribute to eliminating injustice and overcoming the problem of the other."

Happiness arises from awareness. We cannot change the world with a heavy heart; it is essential to face this moment with joy. Because initiating the transformation and seeing that new behaviors make us freer and more decisive is a considerable element. This is why the question concerns not only believers or non-believers but everyone. The path we must follow is one of sharing and defending the Gift, given to us by our fathers. The etymology of the word "community" is this: sharing the *monus*, the gift. Sharing is a positive element that we must put into practice and approach with joy and determination.

Do you know where I met men and women who truly claim to feel at peace with themselves? Among small-scale farmers, the protagonists of food communities. This immediately sidelines the dimension of profit, of money for its own sake. We will not find peace until we search for it in the regeneration of our relationship with the Earth or with food. By losing a healthy and respectful relationship with what we eat, we have lost the meaning of the Earth. We have found it instead a wicked stepmother because we are no longer able to appreciate its good and healthy fruits. We must restart from the concept of care: for the Earth, for food, for people. Let's take an example from the food communities, adopting a lifestyle that can be practiced anywhere in the world. Let's reclaim our reality. We'll find out that it doesn't take any effort. But above all, we will find out that all the energy we invest will be returned to us in the pleasure of living.

G.G. Freedom is being able to contribute to the common good of my community. I am free when I can do that. The liberal freedom to do what one wants is indifferent; it is the freedom to shop at the supermarket or buy from Amazon. The freedom that we imagine is instead the freedom to take an active role in a collective project and commitment.

I'll try to take this reasoning from another, less philosophical point of view. The digital economy is based on the mining industry. Copper, for example, is critical. Studies tell us that in around 2060, we will reach the peak of copper extraction.[2] From that moment on, the availability of this mineral will start to decline. The result? There won't be enough copper for everyone. We will then have a global elite who can access minerals and will be increasingly digitalized, while the rest of the world will live in a condition of backwardness. A dangerous and unjust imbalance. Given their wealth of mineral resources, China and the United States will be on the side of the rich. And Europe or Japan? Being poor from this point of view, they will have to develop a systematic plan for recycling mineral resources.

S.A. Could they do it?

G.G. Under current conditions, it wouldn't be possible because the products we find on the market, from cell phones to cars to all kinds of digital devices, are too complex and therefore too costly to recycle in terms of water and energy consumption. There is much talk about the new generation of electric cars, for example. But if we build them full of microprocessors and electronic particles, from the navigator downwards, we won't have a way to recycle anything. Modern electric cars will need to be very simple and sparse, made of easily recyclable elements. Of course, every time I present these arguments to industrialists, it's a shock for them.

C.P. Here we return to a focal point of our reasoning, the finiteness of resources. It's a principle that struggles to gain acceptance, a concept that we have not yet succeeded in making popular. When Serge Latouche introduced the idea of degrowth, which someone here in Italy has dubbed "happy degrowth," it attracted an impressive amount of attacks and controversy. The crux, in my opinion, is this: Degrowth doesn't mean becoming poorer but, on the contrary, setting a limit to the model of endless capitalism. At the same time, it's necessary to understand what we mean by growth: If it's the growth of our social life, our happiness, our well-being, it's worth pursuing. But if it's only measured by GDP, production and consumption, it won't be genuine progress.

G.G. What you're proposing is a crucial argument. We need a degrowth in the material impact of our lifestyle, production model and consumption habits, while we need growth in our immaterial needs for human and natural relationships. This has nothing to do with poverty. The goal of an individual and a community is not to have more things, in other words to increase GDP. The goal is to live better, to elevate the quality of existence and to find meaning in one's human experience.

CHAPTER 13

People and Nature, A Community of Destiny

S.A. To introduce this section, I turn again to a quote from the French sociologist and philosopher Edgar Morin, who in *Voler bene alla terra* ("Loving the Earth") writes: "Human beings are subjected to a continuous conflict between the pleasure principle and the reality principle, between their need to respect reality and their tendency to deny it. Studying only one of these two aspects is reductionist, as is the dominant science today. The emotional part, with myths and illusions, is fundamental because it does not deny reality but makes it bearable."[1]

In the same work, the Spanish writer Ignacio Ramonet adds: "In the 1960s, many countries thought they could do without farmers, to the point that today there are nations, such as the United Kingdom, that have accepted the idea of living without an agricultural system. The few remaining farmers there, 2% of the population, are subsidized to maintain the landscape, not to actually produce food. Because food can be bought cheaply from the global South! Industrialization wants cloning and GMOs and has even reached the point of producing meat without animals. There is this idea of farming without soil, in the laboratory.

And it's already happening in some cases. There is an attempt to separate agriculture from the land, but farmers know well that the disappearance of the soil means the disappearance of culture. Trees, biodiversity, food identity, tradition in ethnographic terms, songs and superstitions that have deeply marked a local area, everything related to people, festivals, pleasure, the relationship with nature. The culture of a country disappears with the soil. Food is culture."[2]

C.P. Confining biodiversity to a single animal species in the African savannah or to a particular lichen in the Asian tundra means falling into the same mistake made so far: limiting this concept to beautiful, natural, protected oases, marginalizing it from our daily lives and the spiritual sphere. This profoundly flawed approach can still be seen today: a vision born from the anthropocentric conception that envisions nature as a concept separate from human culture.

And so this is why, aware of our dangerousness, in order to safeguard the beautiful and the good, the pure and the immaculate, we tend to keep them away from our actions and thoughts. Nothing could be more wrong! If we truly want to protect biodiversity, the first step must be to reconnect the sphere of culture to that of nature; this implies paying more attention to the biodiversity that, in its many forms, every life on this planet comes across daily.

Also in this case, the key to the problem is once again the limits that with wholly inadequate self-legitimization we tend to impose on nature. Yet the culture that all indigenous communities have developed over the centuries is strongly related to nature and its own sense of limit. This is authentic knowledge respectful of a conception whereby an ecosystem is stronger and healthier the more biodiversity lives in it.

Selecting particular animal and plant species for traits that apparently seem stronger and more resistant, thus reducing the number of species present in our local areas, means undermining our own survival. Recent history has proven this. In the mid-19th century in Ireland the parasitic disease late blight struck the only variety of tuber cultivated on the island, a variety that was the staple of the diet of the very poorest. In just a few years, a million and a half Irish people died of malnutrition and the same number emigrated to the United States. The Irish famine is exemplary of the risk we run when we stop cultivating biodiversity.

What will happen when climate change or unforeseen diseases affect the few species we have entrusted with the task of feeding humanity? We will have no alternative to hunger and famine. We can clearly see how biodiversity, especially related to food, is not something to put on display in a museum or to confine to a protected natural reserve, but a fundamental guarantee for life, health and our food security.

Protecting it means drawing on traditional knowledge and the history of our civilizations. If, in Ireland, to cope with the 18th-century population boom—one of the first effects of the Industrial Revolution—they had not limited themselves to cultivating a single variety of potato but had adopted multiple varieties, as peasants do in the Andes (the tuber's place of origin), many lives, much suffering and extensive migration flows would have been spared. Even today, after centuries of continuous interaction, Andean farmers are used to cultivating up to 32 potato species, which vary depending on the exposure, altitude and micro-habitat of individual plots. This logic probably does not lead to the maximization of profits, but it does protect populations from the risk of famine due to empirical knowledge and daily contact with local biodiversity.

CHAPTER 14

Laudato si', a Universal Document

S.A. Is it possible to imagine building popular sentiment without it degenerating into an absolutist ideology?

G.G. To answer, I'll start with a very personal example. When I tell people that I'm a vegetarian, I often sense a cultural resistance. This is typically embodied by the Western male, well-educated, between 50 and 60 years old, residing in a city, well established professionally and with a medium-to-high salary, who tends to tie his virility to the consumption of meat. If we expand the discussion, we can find an intrinsic relationship between domestic violence, machismo and ecological destruction, with the underlying idea that the market generates wealth and thus solves everything. This is a purely Western predisposition that developed starting from the 17th century. In his radical metaphysical loneliness, the Western man conceives nature as something inert, without a soul, without the Holy Spirit. When we build a road through a forest, we annihilate an ecosystem of relationships among living species that we then don't know how to repair. Trees communicate with each other through a network of fungi that extends below the ground. From the indigenous Amazonian people mentioned by Petrini we must learn to return to taking care of our trees, without which we cannot live.

In the global South countries there is a much more sober and healthy relationship with nature than what we have created here in the West. We are victims of a naturalistic logic in the sense given by the anthropologist Philippe Descola: We believe that everything connecting us to nature is something physical because our bodies are made of the same atoms as the Earth. Consciousness, interiority, the soul or the spirit are instead exclusive to the human species. But is the non-human really without spirituality?[1]

What matters in our lives is not the number of cars we own but the quality of the relationships we have between ourselves and with nature, with our ancestors and our future children. In the language of the Shona people of Zimbabwe, this is called *Ukama*. Trees emit beneficial scents for our nervous system. We are part of the arboreal world. Human society is a complex metabolism in a position of interdependence with other metabolisms: trees, fungi, earthworms, insects, birds. We need to immerse ourselves in this condition and acknowledge it, as many communities do in the global South.

We are facing a great change that is also anthropological, on which Pope Francis strongly insists. I believe we can say that we Europeans, we Westerners, inherit a kind of anthropology symbolized by Leonardo da Vinci's Vitruvian Man. His arms and legs are spread inside a square set inside a circle. If you reflect for a moment, it's the symbol of humanity from the Renaissance onwards. It's a male figure; females are absent. It's a white man; there are no other ethnicities. He is in total metaphysical solitude; nature is absent. He is armed with geometry to dominate the world. There are no children, no elderly, no sick. This is the fantasy that has inhabited our consciousness for four centuries, and it wants nature to be a kind of inert object at the disposal of humanity.

This anthropology, given that it has nothing to do with Christianity, is the same anthropology that today is killing us. Therefore, we need the radical transformation of moving from the anthropology of Vitruvian Man to a relational anthropology, a relational cosmology, in which the relationship is a constituent part. Any two subjects determine themselves based on their relationship. It's not about two distinct entities that come first, and then the relationship. Think also about the relationship with nature and the ancestral relationship between the past and the future.

When we deny these relationships, we throw humanity into the abyss. Then the crisis of ecology becomes the crisis of democracy, and so we need to work to rebuild a new bond between humans and with nature. The issue of food is central to this anthropology that needs to be implemented.

C.P. We are victims of an anthropocentric logic. To contain it, we need to recover spirituality. And I say this as an agnostic. The secular world made a historical mistake by leaving the issue solely to believers, as if spirituality were the exclusive domain of religiosity. Instead, it is a human matter. I, as a secular person who does not believe in a divine dimension, must still realize that nature and animals have rights. I cannot tolerate animals being raised intensively solely for slaughter. I cannot tolerate the disrespect of ecosystems. Because this is not only contrary to logic and health, it is also contrary to my spirituality, regardless of what scientists say, especially in the field of agriculture. Think about biodynamics: Science doesn't explain everything, and the relationship between the times and ways of nature is crucial from the perspective of production. Biodynamics fits into a long tradition of farmers who engage in their practices, even from a more immaterial, almost religious perspective. However, this is not about propitiatory or thanksgiving rituals. At its core is respect and care for the vital systems they are called to safeguard, both the visible ones and the deeper ones that ensure soil fertility. Entrusting oneself to agricultural techniques rooted in the knowledge that farming communities have developed and preserved for centuries requires great courage, especially in a technocratic society like ours. But it also means approaching agriculture with a non-reductionist view, more attentive to the complexity of relationships that regulate nature and, therefore, more far-sighted. Without spirituality, nothing can be accomplished in the fields.

G.G. I agree. I say this as a Christian and a priest: We do not have a monopoly on spirituality. Nor on ecological transition, which is also an anthropological and spiritual transition. We have the task of keeping things united, not dividing them.

C.P. In this sense, the 2019 Synod of Latin America marked a decisive moment: For the first time, an institution like the Catholic Church recognized the ancestral spirituality of the indigenous people, and the Pope invited Christians to understand this spirituality, deeply connected to nature.

G.G. We are at the eclipse of *homo economicus*. We need to build a new human anthropology.

C.P. This is the essence of *Laudato si'* and *Fratelli tutti*, two fundamental documents of Pope Francis's pontificate: You cannot solve the environmental problem unless you work for universal fraternity. If we considered this principle as exclusively Catholic, we would be making a big mistake. This is the seed of a new humanism, the basis of my founding the movement of Laudato si' communities with Domenico Pompili, the former bishop of Rieti and now bishop of Verona.

G.G. The experience of the Laudato si' communities is very important: We need to create and multiply meeting places. And moments of exchange and sharing. The moment is too important for someone to say, "I'll work on it with my association," even if it's a large and important association. We need to organize and be inclusive; Catholics and non-Catholics must stand on the same side in this fight. An open and dialoguing Church must be able to accompany them on this journey.

C.P. Efforts of openness are also needed on the secular side. When *Laudato si'* was released in May 2015, many interpreted it as the Catholic Church's approach to ecology. Anyone who thought that, and still thinks so, shows that they have not understood the scope of that document. Pope Francis's contribution goes beyond an ecological perspective in the strict sense to cross over into a social vision, I would say of social justice. *Laudato si'* gives everyone, whether secular or Catholic, believer or not, an opportunity to change the world. Far from being a purely spiritual matter, it's time to wake up and act! That being said, we thinkers and activists are not here to spread truths or point the way; we must be the expression of a world on the move, of a people marching in need of support and recognition.

Let's go back to Fridays for Future, a breath of fresh air, a very interesting movement though not yet fully defined. As it has only just been established, it wouldn't be reasonable to expect that yet. However, if I make a comparison between them and "my" '68, I would say that this new generation is decidedly more advanced. We were a protest movement and in many ways utopian; in the end, we wanted to achieve one thing, revolution, an impossible goal. Instead, these young people have their feet firmly planted on the ground, they want to change things, but they want to do it (and they do) based on their lived experience: They travel by train, separate their trash, study and bring sustainability principles into the workplace. In short, they put themselves into play in their daily lives, something we never really did.

G.G. The movement has already begun, and it has certain characteristics. This is very true, and this is Jesus's attitude. He came to Earth not to say, "I am the Son of God." He said, "Your faith has saved you."

S.A. How does this approach come about in your experience? Can we say that it is in some way the result of your personal human and spiritual transition?

G.G. The most important part of my life was my experience of civil service in Chad. There I worked as a math and physics teacher in prisons and with street children. In Chadian prisons, men and women are not separated. So every night there were episodes of violence against women, who often became pregnant. The children who survived the terrible hygienic conditions continued to live in prison with their mothers or ended up on the streets. Apart from some clothes, they had nothing, absolutely nothing. Never as in the period during which I was with them did I understand what the joy of living was. At that moment, I decided that I would become a Jesuit to continue to have the opportunity to be with them whenever I could.

A few years later, at the age of 31, I had the chance to work as a trader on Wall Street, at the heart of global finance. An economically significant opportunity. I said no in ten minutes. If I had lived that kind of life, I wouldn't have been able to play with my Chadian street children friends in heaven. They would have told me: "We don't recognize you, we don't know you anymore." Instead, I want to play with the street children for eternity: This is the kind of spirituality that led me to the positions we are discussing today.

The only way to survive and not resemble the people we are fighting against is to maintain a friendship with the poorest. To avoid believing that we are leading a spiritual army, we must give all our energy and creativity to the fight, always remaining detached and free from the struggle.

Now I work at Georgetown University in Washington, DC, and volunteer once a week at a homeless shelter run by the Jesuits. My presence is certainly more useful to me than to those young people, because it allows me to remain human. For me, young people are an incredible source of hope. Another source of great joy is my students from all over the world. I see that they have understood that they are at a historic turning point: They have accepted the principle of ecological transition and have grasped that it is a radical change. They feel the responsibility for a social transformation that humanity has never undergone in the past. Some, initially, struggle with feeling an excess of responsibility and fall into depression, but then experience a resurrection, starting again together, never alone, always in a group.

C.P. I was part of the generation of '68, as a militant in the Italian left. I remember that in 1974, together with a group of friends, we decided to publish the periodical *In campo rosso* ("On a Red Field") with the aim of creating a counterweight on the radical left, which at the time felt stifled by the collaboration between the Italian Communist Party (PCI) and the Christian Democrats (DC). The following year, another initiative started: Radio Bra Onde Rosse, the first private radio in Italy—at the time still under a monopoly regime—which not only broadcast songs but also programs opposing the local establishment. Some legal troubles that led to the seizure of the radio station helped pave the way for the liberalization of independent broadcasts in Italy.

At a certain point, however, my interest focused on the vocation of my area. Bra, in the province of Cuneo, is on the border with the lower Langa, an area with a strong aptitude for winemaking. I went on a pilgrimage to Burgundy and Bordeaux to study the French approach to enology, an approach that did not yet exist in Italy. From there, I started working with small producers. The Slow Food movement, the University of Gastronomic Sciences in Pollenzo and Terra Madre, which brings together 6,000 producers from 150 countries every two years, all developed out of this process.

S.A. Both of you have mentioned it several times: How important was *Laudato si'* in the construction of your thoughts?

C.P. First of all, it must be said that initially, this encyclical struggled to become the subject of deep discussion, both in the secular and the Catholic worlds. Laypeople mostly thought it was the Church's belated line on environmental issues, while the ecological movements already had their roots and history.

In reality, the relevance of *Laudato si'* is extraordinary. Although it was published in 2015, it contains a working hypothesis that still engages readers in good practices and deep analysis. It's not a green encyclical, but a comprehensive social encyclical. Certainly it belongs to the Church's heritage of thinking, but more broadly it addresses all of humanity. The more time passes, the more this reflection becomes important in civil society. In recent years, concern about environmental issues has become more evident, not just trying to understand the extent to which climate change affects crop changes and the lives of millions but also through the mobilization of young people, which was not foreseen when the encyclical was released. The message of *Laudato si'* is more relevant than ever, and its key concept, integral ecology, is the most extraordinary content for the world that awaits us. I believe it is a good tool for reflection and, at the same time, a stimulus for shared practices to face the great season, the great moment of ecological transition that will involve several generations. This is not a process that will be resolved in a few years; it will be long and must have solid foundations, and *Laudato si'* is part of them.

G.G. My impression is that in our Church, the encyclical has had more of an effect on the people of God than on the clergy. Among the faithful, many groups of Christians have gathered to read and meditate on it. The encyclical has also had a considerable impact outside circles of practicing Catholics. It has evangelized many people of good will, previously discouraged by the Catholic magisterium's obsessive insistence on sexual morality. At a time when the Church was a marginal institution in the international community—and I can testify to this as the former chief economist of the French Development Agency—the encyclical pushed our Church to become a kind of locomotive of history. And the pandemic confirmed this. Pope Francis remains the only head of state to have taken a responsible and humane approach during the pandemic.

In any case, *Laudato si'* has helped many become aware that the Gospel is not a private matter. Following Christ involves participating in public affairs and having a relationship with nature and politics. There is a deep unity between caring for creation, respecting children and the elderly and rejecting the illusion of the alleged virtues of the market. This is what paragraph 123 of the encyclical says. It's very strong. Obviously, some who work in finance pretend to embrace the encyclical when, in reality, they do everything to deny it in daily practice. Moreover, among some Christians, there persists great ignorance about the severity of the ecological crisis. Due to certain misinformation, they continue to believe that the Pope is exaggerating.

When it comes to the clergy, things seem complicated to me. Many pastors quickly embraced the cause of the Common Home proposed by the encyclical and helped the faithful to read it. Many others, however, remained hermetic for a long time. It was too far from their culture. These people need to be encouraged to read all of Francis's writings, because they show a strong unity. *Evangelii gaudium* is a great treasure of pastoral wisdom. *Laudato si'* is inseparable from *Fratelli tutti* and *Querida Amazonia*. To enter *Laudato si'*, I believe one must accept entering Francis's entire process, aimed at a profound renewal of the Church.

S.A. How did the Laudato si' communities originate, and how are they increasingly engaged in promoting awareness on the theme of caring for the Common Home?

C.P. The genesis is an intellectual encounter that later grew into a friendship between a "pious agnostic" and a leading Monsignor of the Catholic Church, Domenico Pompili. This was also influenced by the idea of revitalizing Amatrice from within. The town had been devastated by the earthquake that in 2016 struck the province of Rieti, where Pompili was the bishop until 2022, a center that subsequently became a working hypothesis for the future and for reflection on the planet and the local economy. The meeting gave rise to the need to live this experience under the guidance of *Laudato si'*. It immediately occurred to me that the encyclical could be a starting point for a powerful dialogue between believers and non-believers, but most importantly a reflection for future generations. Tools for reflection are crucial when it comes to the ecological transition.

At the same time, the first communities named after the encyclical were established. Without taking a top-down or hyper-associative approach, we saw them slowly but progressively increase. There is a foundational idea expressed in the encyclical which has all the characteristics to go down in history. At the same time, reflection can take place on issues that are of interest not only to the Catholic world but also to the secular world and other religious communities, in an approach that structurally relies on communities.

I like to see how this initiative and the interest it has recently received are steadily growing, just as happened with the reading of *Laudato si'* and its recognition. The task of the communities simply becomes that of being outposts in their local areas, stimulating reflections, discussions and good practices. From this perspective, it is comforting to see a very diverse participation.

G.G. It seems fantastic to me, it seems like the future of Italy, of Europe. We need to work at the political and macroeconomic levels to change structures but we also need to start from the grassroots, as we have already seen with young people. We're talking about an existential experience for everyone: The ecological transition is a spiritual experience and cannot be separated from the people. The project is constituted at the macroeconomic, political and basic levels. Herein lies the difficulty.

S.A. We have discussed the historical importance of *Laudato si'*. However, both of you have a personal relationship with Pope Francis. Can you share some details about your relationship with the pontiff?

G.G. When I met him for the first time, he asked me, "Are you a Jesuit?" I replied yes, and he smiled, saying, "So the Jesuits really are everywhere!" What struck me most about our meeting was the discussion we had about paragraph 123 of the encyclical. The Pope states that the same logic lies behind the mistreatment of children (thinking of the sexual abuse of minors) and the elderly (thinking of those who abandon them) as well as those who think that destroying the environment, biodiversity and the planet is not important. Everything is linked by the same logic. There is a fourth element: Men who abuse women. If we are not even capable of honoring women in our society, then we cannot honor our children, our elderly and our planet. Today, for the first time in Western history, women, on average, are more educated than men. They are slowly gaining power—and it is normal, studying longer—but some men do not understand and resist, including males who hold important roles within our Church. We will never solve the ecological crisis if we do not first resolve the relationship with the acquisition of power by women in the West.

C.P. After 2013, a deep friendship with Pope Francis developed. There were two particularly significant moments. The first, as we mentioned, coincided with the foundation of the Laudato si' communities, together with Monsignor Domenico Pompili. We were faced with the extraordinary intuition of integral ecology, which led to a profound reflection on the necessary changes. It is noteworthy how the subsequent document, *Fratelli tutti*, refers to fraternity, which is the condition for bringing about change. Between the two encyclicals, we see a deep connection: In *Laudato si'*, there is a reflection on the value of dialogue, which can take shape only through fraternity.

The second moment concerns a synodal experience. During a conversation, Francis told me to leave him my address because I would be invited to participate in the special Synod for the Pan-Amazon region. It was a meeting with all the bishops from nine countries with dioceses in the Amazon regions. Despite telling him I was an agnostic, he invited me because "you are a pious agnostic, in the sense that you feel *pietas* for nature."

During the Synod, I experienced what a frontier Church is like: bishops committed to defending indigenous people, their communities and the natural environment against the violence of predatory capitalism. Incredible. The indigenous people danced, played, took the cardinals by the hand and had them dance. I was impressed by both the organization and the content. Each intervention lasted four minutes and was followed by an equal period of silence. The issues were strongly linked to environmental concerns but also to ecclesiastical organization. Some dioceses are as large as all of Northern Italy, and there might be years and years between pastoral visits to a parish. The need for Catholic communities to train new pastors, in other words consecrating village leaders so they could confer the sacraments, was real and urgent. The proposal was put to a vote, and out of 280, 237 voted in favor.

The other great surprise of the Synod, tying back to

Gaël's discussion of the man-woman relationship, was the role of women, both religious and indigenous. Every intervention ended with the urging, "Holy Father, it is time!" The Church has unresolved issues regarding women, impossible to resolve immediately due to internal tensions. This must be kept in mind. At a certain point, however, during the Synod, an indigenous woman, a doctor of anthropology, made a somewhat sensational intervention. She argued that while great strides were being made on the ancestral spirituality of the indigenous people, the issue of women had essentially not yet been addressed. She concluded by saying that it was important to remember that for everyone present, bishops, the Pope, everyone, when they came into the world, the first thing they saw was not a river or a forest, but a woman's vagina. Hearing words like these in a gathering in front of the pontiff is a sign of how we are in the midst of a historically significant change.

CONCLUSIONS

C.P. This is the time of choices. We live in a globalized world that readily accepts a company manager earning a thousand times more than a teacher, and a top-tier footballer earning a thousand times more than a farmer. Ultimately, it seems normal to us because we are immersed in an ideology that makes us believe that these are necessary steps on the road to progress. But an opening is appearing; we are beginning to understand that the pursuit of this ideal is undermining the common home and the system of relationships and creating inequalities that make us fearful and nervous. This is not what we want.

But here comes the emergence of *homo communitarius*, a different conception of humanity, oriented towards solidarity and not competition. It envisions the most vulnerable setting the pace, not the wealthiest, with relationships at the center of existence and essential goods guaranteed to all and managed as common assets. In this framework, common goods and relational goods are the cornerstones of a new organizational and power structure, putting an end to the stranglehold of financial markets. Embarking on this path is possible. It is not a chimera; certainly we need to accelerate and also change the narrative model, but it can and must be done.

Here, we need to return to a point we have already come across in our conversation. Do we want a world with more or less solidarity? Do we want cooperative or competitive citizens? Do we want a population driven by solidarity or individualism? A purely technocratic system or one capable of cultivating spirituality? And, even more profoundly: Do we want our children to grow up in a society where the poor are forever poorer and the rich richer? A society where competition, often unfair, is the sole reason for our economic and social actions? These fundamental questions also represent the political issues that this book wants to push us to address, offering concrete and tangible answers. Faced with the brainwashing of competitiveness, we must play the card of communities of destiny. This weakens the concept of competitiveness, replacing it with that emotional security capable of offering solidarity-based responses to the social anxiety of economic performance at the cost of everything and everyone.

Emotional intelligence rewrites the rules of societies dominated by consumerism and a false semblance of rationality that is, in reality, an unconscious inclination toward self-destruction. On the contrary, the community identifies itself in a common project, in a network of neighboring relationships that generate happiness and balance between humans and nature: This is where we want to go.

G.G. The Christian tradition is a tradition of common goods. At the beginning of the Acts of the Apostles, there are three descriptions of the early Church, where we understand how everything inside that community was a common good. In the fifth chapter, however, we come to the story of Ananias and Sapphira, a property-owning bourgeois couple. The husband, Ananias, decides to sell a piece of property but, in agreement with his wife Sapphira, secretly keeps a portion of the sale proceeds for himself before donating the rest to the apostles. This violates the norm of full communion of goods that governs the Church of Jerusalem. When Saint Peter realizes this, he reacts vehemently, accusing Ananias of lying to the Holy Spirit, recognizing in the deceit the power of Satan. Terrifyingly, the condemnation leads to the sudden death of Ananias. The same fate awaits Sapphira, who, unaware of what happened, appears three hours later, repeating the same version, and receives the same condemnation, dying at Peter's feet. The guilt of Ananias and Sapphira is wanting to privatize a common good. And it is exactly the same "sin" of which businesses like Airbnb or BlaBlaCar are guilty, a sin which has led to neoliberalism taking possession of our lives.

The description of the early Church is a description of a community where everything is organized as a common good. This is not by chance. Christ during the Eucharist tells the apostles who were with him: take and eat, this is my body, it is a common good, shared with the people. And he does the same with his blood. Today, we have the opportunity to reinterpret the Eucharist as a liturgy in which God gives himself as a common good for all of us. And grace is a common good, not a private good, not a public good, not a tribal good, because it is universal.

The story of Cain and Abel is also emblematic. Cain is jealous of Abel because he thinks that God's love can be a private good and, since it has been granted to Abel, this means that he will be deprived of it. He does not understand that God's love is like maternal love, a common good for the whole family. Christianity is the religion of the love of God as a common good. It seems to me that this is what Pope Francis is trying to make clear in the Church. Synodality is a tool to emphasize that decisions must be made within a broader and more articulated debate. This is the synodality that was already the way of operating in the early Church. In the 11th century, Gregory VII revolutionized it, organizing it as a state. At the top was the Pope, an absolute monarch by divine right, served by a bureaucratic class, the clergy, with a universal language, Latin, and a regulatory system, canon law. This model was followed by all the countries of Western Europe. Pope Francis is trying to close this chapter of our history to return to the initial intuition of the early Church, a community where each baptized person takes on a leading role, becoming a missionary and active participant, and ecclesial construction transforms into an inverted pyramid, with the holy people of God as the protagonists in all their diverse forms.

NOTES

Part I: THE END OF A WORLD

1. *We Can't Go On Like This*

1. These are personal "guesstimates" by Gaël Giraud. Most existing climate models do not currently account for carbon released from permafrost in their simulations.
2. Andrew Maddocks, Robert Samuel Young, and Paul Reig, "Ranking the World's Most Water-Stressed Countries in 2040," World Resources Institute, August, 26 2015, https://www.wri.org/insights/ranking-worlds-most-water-stressed-countries-2040.
3. The study, *Changements climatiques dans le massif du Mont-Blanc et impacts sur les activités humaines. Rapport I: Evolutions des scénarios climatiques.* AdaptMB 2019. (CREA Mont-Blanc et l'ARPA Vallée d'Aoste, Janvier, 2019), predicts that there should be an increase of the average temperature around the Mont-Blanc between 2 and 3C. It should let the isothermic threshold (above which temperatures never exceed 0C) rise by 300m, so as to reach the altitude of 4100m. A second study, Hillary Gerardi, "Rapport AdaPT Mont-Blanc: quand la montagne sera verte," Blog Crea Mont Blanc, February 10, 2020 https://blog.creamontblanc.org/?p=2619,

suggests that the altitude at which forests can survive would rise by 100m in 2050. These factors will presumably lead to a greening of the Mont-Blanc.

4. See the UN report at http://esa.un.org/unpd/wpp.
5. "Nuovo rapporto IDMC: sono 60,9 milioni gli sfollati interni nel mondo," OIM UN Migration, May 11, 2023, https://italy.iom.int/it/news/nuovo-rapporto-idmc-sono-609-milioni-gli-sfollati-interni-nel-mondo#:~:text=La%20Banca%20Mondiale%20ha%20previsto,non%20verranno%20adottate%20misure%20efficaci.
6. Emanuele Bompan and Marirosa Iannelli, *Water Grabbing* (EMI, 2018).
7. "High and Dry: Climate Change, Water, and the Economy," World Bank Group, May 2, 2016, https://www.worldbank.org/en/topic/water/publication/high-and-dry-climate-change-water-and-the-economy, states that water scarcity, exacerbated by climate change, could cost some regions up to 6% of their GDP. "Ebb and Flow: Water, Migration, and Development," World Bank Group, August 23, 2021, https://www.worldbank.org/en/topic/water/publication/ebb-and-flow-water-migration-and-development, indicates that water deficits are linked to 10% of the increase in total migration within countries from 1970 to 2000.

2. *Too Much Food and Too Much Hunger*

1. Hannah Ritchie, Pablo Rosado, and Max Roser, "Breakdown of carbon dioxide, methane and nitrous oxide emissions by sector," Our World in Data, last revised January 2024, https://ourworldindata.org/emissions-by-sector; "FAO: I sistemi alimentari contribuiscono per oltre un terzo alle emissioni

mondiali di gas a effetto serra," United Nations, March 9, 2021, https://unric.org/it/fao-i-sistemi-alimentari-contribuiscono-per-oltre-un-terzo-alle-emissioni-mondiali-di-gas-a-effetto-serra/.

2. "Fao, mentre cresce la fame il 30% del cibo viene sprecato nel mondo," ANSA, September 27, 2023, https://www.ansa.it/canale_terraegusto/notizie/istituzioni/2023/09/26/fao-mentre-cresce-la-fame-il-30-del-cibo-sprecato-nel-mondo_d65a507c-c299-4b65-8893-7486c76bc2ee.html.
3. "Rome Water Dialogue 2023 seeks solutions and synergies and brings practical contributions," Food and Agricultural Organization of the United Nations, October 4, 2023, https://www.fao.org/newsroom/detail/rome-water-dialogue-2023-seeks-solutions-and-synergies-and-brings-practical-contributions/en; "Acqua nel mondo: quello che c'è da sapere sull'Oro Blu," Acqua del Rubinetto, September 24, 2021, https://acquadelrubinetto.gruppocap.it/ambiente/acqua-nel-mondo/.
4. "2024 Global Report on Food Crises," Global Network Against Food Crises, https://www.fsinplatform.org/report/global-report-food-crises-2024/.
5. "Più di un miliardo di persone nel mondo convivono con l'obesità," Consiglio Nazionale delle Ricerche, March 4, 2024, https://www.cnr.it/it/nota-stampa/n-12566/piu-di-un-miliardo-di-persone-nel-mondo-convivono-con-l-obesita.
6. "La fame in cifre," Food and Agriculture Organization of the United Nations, https://www.fao.org/interactive/state-of-food-security-nutrition/2023/it/.
7. Daniele Banfi, "Ecco perchè e come nasce il vizio del fumo," Fonazione Veronesi, March 23, 2011, https://www.fondazioneveronesi.it/magazine/articoli/oncologia/ecco-perche-e-come-nasce-il-vizio-del-fumo-2.

8. "Rapporto ONU: salite a 828 milioni nel 2021 le persone che soffrono la fame nel mondo," Food and Agriculture Organization of the United Nations, July 6, 2022, https://www.fao.org/newsroom/detail/un-report-global-hunger-SOFI-2022-FAO/it.
9. "Puoi sconfiggere il sistema? Gioco di simulazione," Oxfam Italia, https://www.oxfamedu.it/wp-content/uploads/2020/04/puoi_sconfiggere_il_sistema_descrizione_attivita_0.pdf.

3. *A Planet of Plastic*

1. Silvio Greco, *La plastica nel piatto, Quando e come siamo diventati plasticofagi* (Giunti-Slow Food Editore, 2020), 30–31.
2. Greco, *La plastica nel piatto,* 31.
3. Greco, *La plastica nel piatto,* 31.
4. Riccardo Barlaam, "Microplastiche: ecco quante ne ingeriamo. E la quantità è destinata a triplicare," 24 Mondo, July 7, 2019, https://www.ilsole24ore.com/art/microplastiche-ecco-quante-ne-ingeriamo-e-quantita-e-destinata-triplicare-AC5Sg2T; Lucio Biancatelli, "Sempre piu' plastica nella nostra dieta," WWF, June 13, 2019, https://www.wwf.it/pandanews/ambiente/sempre-piu-plastica-nella-nostra-dieta/.
5. Nathalie Gontard and Hélène Seingier, *Plastique: le grand emballement* (Éditions Stock, 2021), 106.
6. Gontard and Seingier, *Plastique*, 36.
7. Gontard and Seingier, *Plastique*, 36.

4. *Economics of Suffering*

1. Composer un monde en commun—une théologie politique de l'Anthropocène, coll. "la couleur des idées," Seuil, 2022, 586.

5. *Toxic Finance*

1. Hugo A. Martin, Aurélien Quiquet, T. Nicolas, Gaël Giraud, Sylvie Charbit, et al. "Climate-Induced Economic Damages Can Lead to Private-Debt Tipping Points," HAL Open Science, February 9, 2024, https://hal.science/hal-04224077/document.
2. Gaël Giraud et al, "Fossil assets: the new subprimes? How funding the climate crisis can lead to a financial crisis," Report Rousseau Institute, June 2021, https://reclaimfinance.org/site/wp-content/uploads/2021/06/Report-Fossil-Assets-the-new-subprimes.pdf.
3. These figures reflect personal computations by Gaël Giraud. See also John Komlos, "The Actual U.S. Unemployment Rate in 2019 Was Twice the Official Rate, and the Phillips Curve," *Challenge* 64, 2021, 51–74, DOI: 10.1080/05775132.2020.1863547, where the level of underemployment is found to be roughly 7%; and https://www.shadowstats.com/, where the actual level of underemployment is claimed to be as high as 20%.
4. Source: "GDP (current US$) – Italy," World Bank Group, https://data.worldbank.org/indicator/NY.GDP.MKTP.CD?locations=IT.
5. This figure reflects personal computations by Giraud.
6. According to the 2022 BloombergNEF report (https://about.bnef.com/energy-transition-investment/), at this time, roughly 50% of worldwide investments in green infrastructures and technologies originated in China, making it the first green investor relative to its GDP. Since then, the "build back better" strategy and the "inflation reduction act" (2022) of the Biden administration put the US back in the lead.
7. Lars Osberg and Andrew Sharpe, "An Index of Economic Well-Being for Selected OECD countries," *Review of Income and Wealth Series* 48, no. 3 (2002): 291–316, 10.1111/1475-4991.00056.

8. Gaël Giraud, *La rivoluzione dolce della transizione ecologica* (Libreria Editrice Vaticana, 2022).

6. The Failure of International Institutions

1. Stati per spesa militare," Wikipedia, last modified January 14, 2025, https://it.wikipedia.org/wiki/Stati_per_spesa_militare.
2. Andrea Barolini, "Spese militari, record storico nel 2021: 2.100 miliardi di dollari," Valori, April 26, 2022, https://valori.it/spese-militari-record-storico-2021-sipri/.
3. "Meno del 3% della spesa militare del G7 basterebbe per contribure ad azzerare la fame nel mondo e risolvere la crisi debitoria dei paesi del sud globale," Oxfam, June 12, 2024, https://www.oxfamitalia.org/wp-content/uploads/2024/06/CS_OXFAM_MENO-DEL-3-DELLA-SPESA-MILITARE-DEL-G7-BASTEREBBE-PER-CONTRIBURE-AD-AZZERARE-LA-FAME-NEL-MONDO-E-RISOLVERE-LA-CRISI-DEBITORIA-DEI-PAESI-DEL-SUD-GLOBALE_12_6_2024_definitivo.pdf?_gl=1*1y897ya*_up*MQ..*_ga*MTM3NjUyMjYyMi4xNzMwMzY4MzA1*_ga_1WNLKELH9K*MTczMDM2ODMwNS4xLjEuMTczMDM2ODMyMi40My4wLjA.

Part II: A NEW WORLD

7. First Civil Society, Then Politics

1. "Affluenza, alle regionali 2023 crolla la partecipazione: votano quattro elettori su dieci. Il peggior dato della storia della Repubblica," la Repubblica, February 13, 2023, https://www.repubblica.it/politica/2023/02/13/news/affluenza_elezioni_2023_lazio_lombardia-387805535/.

2. Marco di Dell'Aguzzo, "Ecco i Paesi più esposti al grano ucraino (e non solo)," *Start Magazine*, June 10, 2022, https://www.startmag.it/economia/grano-ucraina-paesi-dipendenti/. See also the Black Sea Grain Initiative figures available at https://www.un.org/sites/un2.un.org/files/bsgi_facts_and_figures_final.pdf.
3. Paolo Mastrolili, "Stati Uniti, la Corte Suprema limita i poteri dell'agenzia ambientale," la Repubblica, June 30, 2022, https://www.repubblica.it/esteri/2022/06/30/news/usa_corte_suprema_sentenza_ambiente-356071455/.
4. Rob Hopkins and Lionel Astruc, *L'ecologia di ogni giorno. Terra, cibo, comunità. La Transizione, un nuovo modo di stare al mondo* (EMI, 2016).

8. *Protectionism and Food Sovereignty*

1. The Rodotà Commission's 2007 bill to introduce the category of common goods into the Italian Civil Code is not publicly available. An account of it can be found in Antonio Vercellone, "L'Expérience des biens communs en Italie—Espaces urbains, propriété privée, droits fondamentaux," Érudit, April 6, 2023, https://id.erudit.org/iderudit/1098401ar.
2. Alain Supiot, "Labour is not a commodity: The content and meaning of work in the twenty-first century," *International Labour Review* 160 (2021): 1–20, https://doi.org/10.1111/ilr.12205.
3. Salvatore Ceccarelli and Stefania Grando, *Seminare il futuro. Perché coltivare la biodiversità?* (Slow Food Editore-Giunti, 2019).
4. Figures are based on Carlo Petrini's personal computations (provided one excepts 2020, where Chinese imports from Africa shrank because of the COVID-19 pandemic). According

to the Ecofin agency, Chinese imports from Africa rose by 14% in the 1st half of 2024. "Hausse de 14% des importations chinoises venant d'Afrique au 1er semestre 2024," Agence Ecofin, July 22, 2024, https://www.agenceecofin.com/actualites/2207-120516-hausse-de-14-des-importations-chinoises-venant-dafrique-au-1er-semestre-2024.

10. Revolution at the Table

1. "La carne: un alimento fondamentale in continua evoluzione," WaFS, accessed November 2024, https://www.waterandfoodsecurity.org/scheda.php?id=142#:~:text=I%20Paesi%20sviluppati%20consumano%20circa,%C3%A8%20inferiore%20ai%2010%20kg.
2. "Carne Coltivata," GFI Europe, accessed November 2024, https://gfieurope.org/it/carne-coltivata/#:~:text=Una%20ricerca%20di%20CE%20Delft,%C2%A34%2C80%20al%20kg.
3. "Spreco alimentare: la FAO sottolinea che il 30% del cibo a livello mondiale viene sprecato o perso," Alimenti&Salute, last modified September 27, 2023.

11. Community Economies

1. Important Update: Continuation of Our Programming," Farmers Rising, accessed November 2024, https://www.farmersrising.org/ao-farm-closing.html.
2. Zygmunt Bauman, *Consuming Life* (Polity Press: 2007), 57, https://realsociology.edublogs.org/files/2013/09/168709399-Zygmunt-Bauman-Consuming-Life-2007-1107cis.pdf.
3. "Road to Net Zero," Institut Rousseau, accessed November 2024, https://institut-rousseau.fr/road-2-net-zero-en/.

4. Domenico De Masi, *Ozio Creativo: Conversazione con Maria Serena Palieri* (Rizzoli, 2019).

Part III: TOWARDS A NEW SPIRITUALITY

12. The Happy Transition

1. Luis Sepúlveda and Carlo Petrini, *Un'idea di felicità* (Slow Food Editore and Ugo Guanda Editore, 2014), 49.
2. Olivier Vidal, Fatma Rostom, Gaël Giraud and Cyril François, "Global Trends in Metal Consumption and Supply: The Raw Material-Energy Nexus," *Elements* 13, no. 5 (2017): 319–24; and Olivier Vidal, Fatma Rostom and Gaël Giraud, "Prey-Predator Long-Term Modelling of Copper Reserves, Production, Recycling, Price and Cost of Production," *Environmental Science and Technology* 53, no. 19 (2019): 11323–11336.

13. People and Nature, a Community of Destiny

1. Edgar Morin, "A True Polity of Planetary Civilization," in *Loving the Earth*, ed. Carlo Petrini (Slow Food Editore, 2014), 30–37,
2. Ignacio Ramonet, "A Different World Is Possible," in *Loving the Earth*, ed. Carlo Petrini (Slow Food Editore, 2014), 118–25

14. Laudato si', a Universal Document

1. Philippe Descola, *Beyond Nature and Culture*, trans. Janet Lloyd (University of Chicago Press, 2013), 633.

ABOUT THE AUTHORS

Gaël Giraud, SJ, is a Jesuit priest and a senior researcher in economics at the French National Research Center (CNRS). An alumnus of the Ecole Normale Supérieure (Paris), he holds a PhD in mathematics from the Ecole Polytechnique and a PhD in theology from the Jesuit university of Paris. Nominated as the best young French economist in 2009 by Le Monde, he has served as chief economist and executive director of the French Development Bank, as well as a professor at the McCourt School of Public Policy and founding director of the Environmental Justice Program, both at Georgetown University. He is currently the president of the Rousseau Institute, a French think tank dedicated to ecology, economics, and politics. His research focuses on quantitative models for macro-economics and climate change. He also serves as an ambassador of the Economy of Francesco.

Carlo Petrini is the founder of the global Slow Food movement (1986) and the University of Gastronomic Sciences (2004), in Pollenzo, the world's first academic institution to offer an interdisciplinary approach to food studies, of which he still serves as president. He is also the creator of Terra Madre, a network of over 2,000 food communities that brings together small-scale farmers, fishers, artisans, chefs, young people, academics and experts from more than 150 countries. In 2008 he was the only Italian included in the *Guardian*'s prestigious list, "Fifty People Who Could Save the World," and in 2016 he was appointed by the president of the Food and Agriculture Organization of the United Nations, Graziano da Silva, as FAO Special Ambassador to Europe for the program "*Fame Zero*" (Zero Hunger)."

Stefano Arduini is the managing director of *Vita* magazine and Vita.it and its former editor in chief. He co-authored, with Maurizio Carrara, *Storie di straordinaria filantropia* ("Stories of Extraordinary Philanthropy"), published in 2011.